FRIENDS
of the
Davenport Public Library

"Celebrate The Printed Word"
Endowment Fund
provided funds for the
purchase of this item

AN
ANATOMY
OF PAIN

*How the Body and the Mind Experience
and Endure Physical Suffering*

Dr. Abdul-Ghaaliq Lalkhen

Scribner

New York   London   Toronto   Sydney   New Delhi

Scribner
An Imprint of Simon & Schuster, Inc.
1230 Avenue of the Americas
New York, NY 10020

First Scribner hardcover edition February 2021

SCRIBNER and design are registered trademarks of The Gale Group, Inc., used under license by Simon & Schuster, Inc., the publisher of this work.

For information about special discounts for bulk purchases, please contact Simon & Schuster Special Sales at 1-866-506-1949 or business@simonandschuster.com.

The Simon & Schuster Speakers Bureau can bring authors to your live event. For more information or to book an event, contact the Simon & Schuster Speakers Bureau at 1-866-248-3049 or visit our website at www.simonspeakers.com.

Interior design by Erika R. Genova

Manufactured in the United States of America

10 9 8 7 6 5 4 3 2 1

Library of Congress Cataloging-in-Publication Data has been applied for.

ISBN 978-1-9821-6098-2
ISBN 978-1-9821-6099-9 (ebook)

*To my wife, Nichola, whose patience and kindness has no equal,
and to my parents and siblings, who have always been both the
anchors of my life and the wind in my sails.*

But pain is perfect misery, the worst of evils, and, excessive, overturns all patience.

—John Milton, *Paradise Lost*

CONTENTS

INTRODUCTION

An Unexpected Journey

There is a special quality to the sound of the telephone when it rings at two in the morning. When you are working as the anesthesiologist on a labor and delivery ward, that call causes your heart to race and your stomach to do a little flip, adding another layer of fatty damage to your overstressed coronary arteries. You have usually just fallen into an uneasy sleep on a lumpy bed in a damp but overheated on-call room, where many doctors have tossed and turned before you. The voice on the other end of the line is typically that of a harassed and overworked midwife. "Epidural—room 4" might be all you hear before the deafening click of the slammed handset. There is the urban legend of an anesthesiologist who, on receiving such a message, walked up to the room he had been sent to and stuck an epidural needle directly into the door before turning around and going back to his bed.

On entering a birthing room at 2:00 a.m., when the human spirit is at its lowest ebb, the scene is often predictable. There is a partner who looks scared out of his wits, a midwife who is fretting and trying her best to reassure everyone, and a pregnant woman who is screaming in pain. You are greeted by an individual who would normally present a reasonable and calm image to the world but who is now reduced to a feral and illogical specter as she is consumed with pain. The pain experienced during labor does not spare social class, color, or creed. Pain reduces us all to our basest elements; it is a great equalizer and unifier and one thing we all experience.

As an anesthesiologist I am confronted by a woman who is demanding pain relief and is now screaming at me and everybody who comes into her line of vision. Administering an epidural in these circumstances is challenging, to say the least. When I return later, however, after the epidural has taken effect, I am greeted by somebody completely different: a calm, personable individual. The ability to transform someone's demeanor completely by abolishing a barrage of unpleasant sensations that are coming from distended and damaged tissues has made a lasting impression on me, and I continue to be amazed at the transformation that takes place when pain is removed.

While we now accept pain relief in labor as being an absolute right—and in first-world countries epidurals are provided on demand—this was not always the case. We have vacillated over the centuries between advocating that pain should be aggressively treated and believing that pain is necessary and important to the curing of the condition or an integral part of the treatment being given. Inflicting suffering was thought to sometimes be inevitable, for example in the case of amputations before the invention of anesthesia or when "releasing evil humors" by bloodletting with leeches. Pain was sometimes thought to be intrinsic to the success of the therapy, a sign of the patient's vigor and helpful to their personal growth. "Pain is weakness leaving the body" was a phrase often shouted by my rugby coach. "Pain is good for you. It tells you that you are still alive" was a lesson sometimes uttered by stressed doctors in the emergency department where I worked, when the umpteenth stab victim of the evening complained about the chest drain inserted to inflate a collapsed lung.

At one time priests believed that the pain of childbirth strengthened the bond between mother and child, reinforcing the idea that self-sacrifice is inherent in motherhood, and that interfering with the pain would disrupt this maternal bond. Even in the early days of medical procedures, pain was not always seen as something that needed to be treated, largely because it was a sign of whether the patient was well and whether the treatment was working. It was only after 1800 that doctors became increasingly preoccupied with managing pain from injury or surgery, treating patients who had painful cancer as well as those with arthritis and migraines. When surgical anesthesia was first introduced in the mid-nineteenth century, however, there was uncertainty about the ethics of operating on an unconscious person and a concern that pain relief might retard the healing process. Priests, and some physicians, feared that anesthesia would corrupt the individual's soul, bringing into play a moral calculator to assess whether or not pain should be relieved. The development of medicine and surgery as endeavors that can be accomplished only by relieving pain, and a philosophical shift toward the value of the individual, changed the narrative from one claiming pain was necessary and a part of human life to viewing pain as an experience that needed to be actively managed.

Religious writers initially deemed anesthesia in the context of childbirth a violation of God's law. This was obviously an idea proposed by men, and it was not until Queen Victoria had chloroform to facilitate the birth of one of her children that the practice of providing women with analgesia during labor achieved acceptance. Even today the application of pain relief in labor varies across the globe; the mechanical stretching and contraction of tissues during labor is interpreted, perceived, and responded to through each woman's unique social and cultural lens. In Chinese and Korean cultures, for example, birth must appear to be pain-free and is endured in silence in order to not bring shame to the woman's family, whereas in some other cultures a more vocal response is believed to ensure a more solicitous and attentive husband.

What is it about pain that makes some people require medical intervention and others not? The storm of information created and transmitted by nerves when we are burned, stabbed, bumped, bruised, shot, frozen, disappointed in love and life, jarred, shaken, and irradiated is perceived, interpreted, and modulated by our individual brains, which are as unique as snowflakes famously are. Pain, too, is therefore unique and individual and often confounds us and those we love and who love us.

━━━━━━

I work at the Manchester and Salford Pain Centre at Salford Royal National Health Service Foundation Trust in the United Kingdom. The hospital is a major trauma and neurosciences center located about three miles from the Manchester United soccer stadium. I split my time between anesthetizing patients in the operating room, inserting electrical devices to relieve chronic pain, and assessing individuals with chronic pain in an outpatient setting. But the reason I chose a career in anesthetics was to become an intensive care doctor.

I was born and raised in South Africa, attending school and university under the shadow of Table Mountain. As a junior doctor I worked in a major trauma center and was exposed to some of the most horrific human-inflicted trauma imaginable. During the nine months I spent in an inner-city trauma unit I developed a deep and pervasive antipathy toward people in general and felt that I would survive in medicine only if my patients were asleep and my interaction with them was brief. I found myself more comfortable with patients who were uncommunicative and the stabilization of the injured person less emotionally taxing than a biopsychosocial assessment of the journey that had brought them to the hospital. It didn't help that at the time I was unhappily in love. This was also when some influential government leaders denied that HIV causes AIDS. When this statement was repeated to me by a patient who was

HIV-positive and refusing to disclose this information to her partner, I coped by running away from home.

I went on to become an anesthesiologist, completing my training in the UK, where I have remained. Over time I specialized in pain medicine, an area often assigned to departments of anesthesiology, for reasons we will come to later. I now work in a clinic with patients who suffer acute pain—that is to say, pain with an easily identifiable cause (traumatic injury or surgery) that will eventually go away when the physical damage heals—or with chronic pain, which may not have a clear cause and may not go away.

The primary emotion a doctor feels when faced with somebody who is in pain is extreme helplessness. In many ways pain is designed to elicit this response from others; we are social animals, and one of the purposes of a visible pain experience, manifest as pain behaviors, is to encourage help from others—pain behaviors have survival value. Seeking assistance after an injury is as biologically imperative today as it was a thousand years ago; what has changed, however, is the place of healing and the people we seek healing from. A thousand years ago a sick person might have gone to a temple to meet with a priest; health and spiritual well-being were inextricably linked, and so the priest served as both healer and spiritual guide. At an even earlier time in human history, the person consulted might have been the local medicine man, who would have communicated with the ancestors to find answers to various afflictions. Today we have huge biomedical temples called hospitals and networks of shamans called general practitioners, organized into a hierarchical system and monitored by professional bodies.

The nature of the interaction between doctor and patient has changed over the centuries based on our accumulated knowledge of physiology and anatomy, as well as our understanding of disease and treatments. Today most individuals can understand traumatic pain without any knowledge of the complex neurophysiology involved. We understand to a degree that if we injure ourselves, then we experience pain, in the same way that a car or house that is broken into results in activation of an alarm. Pain is an alarm bell that something is wrong and a call to seek medical help. In the modern world, however, our understanding of pain as an alarm system has a limited benefit in a society where we voluntarily subject ourselves to the trauma of surgery, for example—and yet pain continues its primitive wail despite our enhanced ability to understand and manage trauma and disease. It is this disconnect between the biological and the psychological that often makes modern pain management ineffective. Even when pain is due to an injury, seeking help is still governed by the individual's perception of the severity of the injury and their health-seeking behavior, which in turn is governed by a host of psychosocial and cultural factors. Most people, including doctors, do not appreciate that the organ that produces pain is the brain. It is not the broken bone or the damaged tissue or the bleeding wound. The experience of pain is the sum total of more than just the physical injury—it is the result of this information being filtered through the individual's psychological makeup, genetics, gender, beliefs, expectations, motivations, and emotional context.

We have a collective delusion that the human body is a simple machine that can be repaired by the medical profession when it breaks down. However, the reality is that the human body is not a simple machine; it is an organism that progressively deteriorates over time and then, eventually, ceases to function. Pain as an experience is often seen simplistically as a manifestation of a dysfunctional machine and is often a consequence, when it becomes chronic, of a machine that has been poorly treated. Unfortunately, rather than recognizing that it is the behavior toward the machine that needs to change, we have a tendency to medicalize this pain and therefore often apply excessive amounts of lubrication in the form of opioids and surgery, which do nothing to improve the function of the machine and in fact can cause a significant deterioration of the organism.

Simply put, we need to stop viewing our bodies as machines that medicine can fix when they go wrong. If people are educated about their body and about pain, perhaps we might put an end to the undulating fashions of medical meddling—meddling that, in some cases, has far-reaching consequences appreciated only decades later. We are currently, for example, at the tail end of an epidemic of opioid prescribing and are beginning to see the effect of that rampant scourge on individuals and on society. Once upon a time opiates were the province of recreational drug users, and abstaining from them, even in the face of severe pain, was considered not only appropriate but admirable. The changing understanding of the possible advantages of treating pain after surgery or injury swung the pendulum toward increased opioid prescribing, with the relief of pain regarded as a human right and a benefit to the patient. We now see addiction, tolerance, dependence, and withdrawal as well as the long-term physical effects of prescription opioid use on patients. As medical professionals, we have begun to realize that our overzealous management of acute postoperative pain and our mistaken treatment of chronic pain (as opposed to acute pain due to injuries) with the same medication has had unintended consequences.

My mission in this book is to explain pain in all its forms: pain from physical trauma, cancer pain, and pain that appears to continue in the absence of any physical damage. We all suffer with pain at some point in our lives and are witness to the pain of those we love and care for. With renewed knowledge and understanding, we can become active participants in the art of caring, understanding, and coping with an experience that can become all-consuming.

CHAPTER ONE

How Does Pain Work?

You may have purchased this book to read while on vacation somewhere, as you bask in sunshine. Perhaps you are being tortured or delighted by the excited screams of children or feeling gentle breezes blowing from the sea, while the smell of sunscreen permeates the air. Your swimsuit may be slightly damp from being in the pool. Regardless of the sounds, smells, and temperatures you are exposed to, these sensations are all quite easy to tune out so they become background noise. However, if you were to be bitten by a mosquito or an exposed part of your body started to object to the strong sunlight, this experience would demand your attention and it would take a great deal of conscious effort to ignore. This is the universal experience of pain. Physiologists refer to pain as "aversive at threshold," which means that it cannot be ignored or subdued easily. By its very nature it demands attention, and this is as true today, as you sit languishing idly around a swimming pool, as it was when we were a primitive species fighting for survival in a harsh and untamed world.

But while damage to our body may eventually demand our attention through the experience of pain, this experience can be ignored, sublimated, or delayed by the brain. Pain is a warning system, informing us that there is a threat to the safety of our body or even that damage has already occurred, but if experiencing pain and receiving this information is not immediately beneficial, then the message relaying this information will be de-prioritized and sometimes ignored by the brain. We all know how crippling and all-consuming the experience of pain can be, and if it might delay, for example, our flight to safety, then it is not immediately helpful and could even be dangerous. The relationship between physical damage and the experience of pain can tell us a lot about the complexity of the biological pain alarm system and the processing of the information about damage, the route this message takes, and why, how, and when it can be disrupted.

While playing for Milan against Chievo on March 14, 2010, the soccer star David Beckham ruptured his Achilles tendon in the eighty-ninth minute of the game. Video footage shows him turning sharply and trying to control the ball; the sharp turn is probably what caused the injury. He then starts to limp because his ankle will no longer flex and extend since he has now lost the use of his calf muscles, which rely on being fixed to the ankle bones via the Achilles tendon. It appears that initially he does not realize he has injured himself. I imagine most professional athletes exist with a level of discomfort that would be abnormal to the rest of us, and so Beckham's natural tendency would be to ignore the messages reaching his brain and carry on in the context of practicing his profession. He tries again to run up to the ball to kick it but finds that he cannot perform this action; his progress is halted not by pain, as yet, but by loss of mechanical function. The tear to his Achilles tendon occurred moments before, and while the process by which information about this damage is converted to an electrical signal began at the moment of injury, it has not yet registered in his brain as pain.

When you sustain an injury, the traumatized tissue releases and attracts chemicals called inflammatory mediators. The aim of these substances is to heal the damaged tissue, but they also play a role in triggering the pain alarm. The release of chemicals such as hydrogen ions, potassium ions, bradykinins, and prostaglandins stimulates the harm-sensing receptors in the tissues. The first step in the body's complex pain alarm system, which we all possess (unless we are born with congenital hypoalgesia, a condition where people do not feel pain), is the conversion of damage to the body into an electrical signal. Throughout our bodies there exist harm-sensing receptors (which are like locks on doors) on free nerve endings called nociceptors; the prefix *noci* means "harm" or "mischief" in Greek. Harm-sensing nerve endings are widely distributed in the skin, muscle, joints, organs, and the lining of the brain and are either covered with myelin, which is a fatty tissue, or are uncovered. Myelin-sheathed nerve endings (A delta fibers) conduct electricity faster than their thinner, uncovered counterparts (C fibers); the faster signals from the A delta fibers make you instantly remove your hand from a hot object, whereas the slower fibers produce the sensation that teaches you not to touch the hot object again.

The human body can be injured in only three ways: by mechanical trauma (such as gunshots, stabs, bumps, and scrapes), chemical injury (a burn from an acid or an alkaline substance), and injury from extremes of temperature. All of these injuries result in the release of inflammatory mediators. (Inflammation can also occur when the body's immune system attacks itself or joints become inflamed.) Nociceptors are of different classes and, like locks, are opened with different keys: intense pressure, temperatures greater than 104 to 113°F or less than 59°F, or chemicals released from injury and inflammation. In Beckham's case, the damaged Achilles tendon released histamine, serotonin, bradykinin, hydrogen ions, and other substances that insert themselves into nociceptors, triggering an electrical impulse that communicates and codes for harm and damage and begins its journey to the brain, where that message can be decoded.

The information that his body had been damaged first passed from Beckham's Achilles tendon to an area of the spinal cord called the dorsal horn; the information then passed into the substance of the spinal cord, which is an extension of the brain and enables communication between the brain and the rest of the body. The dorsal horn is constantly receiving information both from the outer reaches of the body and from the brain via the spinal cord; it is like a bowl of soup whose flavor can be altered by inputs from the brain or from the peripheral nerves, rather than existing as a hardwired, fixed computer component. The face and neck are slightly different from the rest of the body, in that the nociceptive nerve endings meet in a structure called the trigeminal ganglion, which projects to the brainstem that is located in the base of the brain.

Once the thin C fibers and fat A delta fibers have conveyed information to the first and second layers of the dorsal horn, a chemical substance called glutamate is released from the dorsal horn nerves when the electrical signal generated from nociceptor activity reaches it. This opens doors within the spinal cord, sending a message to the brain about what has happened. The graver the injury, or the more times it is inflicted, the more keys are made available and the more doors are opened, resulting in more information being sent from the spinal cord to the brain—a phenomenon called wind-up, which contributes to the persistence of pain even when the injury has healed.

There are five superhighways of information ascending to the brain from the spinal cord: the spinothalamic tract (the "where is the problem" pathway), the spinoreticular and spinomesencephalic tracts (triggering arousal, emotion, fight-or-flight instincts and activating motivation circuits in the brain that determine how we behave toward injury based on previous experience), and the cervicothalamic and spinohypothalamic pathways (controlling the regulation of hormones).

From Beckham's Achilles tendon this information rushed to the brain via these superhighways in the spinal cord and lit up different parts of his brain like the colors from a single firework that spread across the night sky. The information traveled to the area of the brain called the thalamus, which is connected to the sensory part of the brain that deals with location and sensation (spinothalamic), as well as the areas responsible for sensing and coordinating emotion (spinohypothalamic), mediating the emotional components of pain. The survival and awareness areas of the brain are reached via the spinoreticular circuit, bringing into play the areas that facilitate an increase in heart rate, blood glucose, and blood pressure to deal with whatever danger may be around. Connections are also made to the parts of the brain that deal with motivation. We are driven by that which is necessary for survival—food, sleep, the avoidance of pain—and we are also driven by rewards, which can be any experience that facilitates learning or results in pleasure; these motivation circuits develop early on and continue developing throughout our lives, determining how we behave. The damage signals also reach the parts of the brain that can release substances that cause pain relief (naturally occurring opioids and cannabinoids). And connections are made to the pathways that descend from the brain to the spinal cord and are responsible for suppressing signals from the spinal cord, reducing the unpleasant information reaching the brain and thereby reducing the pain we experience. Pain is therefore always a sensory and an emotional experience, as these "superhighways" connect to areas of the brain that involve both.

What is interesting when watching footage of Beckham being injured is that initially there is no indication that he is in pain—he simply frowns, more confused than distressed, and continues to try to play. He eventually realizes that he cannot kick the ball and examines the area of his body that is not working properly. He knows where he has been injured because his brain has already received a message from his Achilles tendon, which has registered in the part of the brain that deals with pain location, but we do not yet see his outward expression of pain because the information is still being considered and evaluated.

What follows his assessment of the injured area is his realization of the injury and, more important, its meaning for him as an individual. This information is processed through the parts of his brain that deal with emotion and context (the hypothalamus). I'll bet his initial thought was "No World Cup. There goes my chance of captaining England." Once this information has been processed by his brain, we see him collapse under the weight of the implication of this injury. He lies on the ground distraught, holding his head in his hands. We see not only the expressions of pain but also the very visible experience of suffering. The sequence of events that we have witnessed, from the moment he struggles to kick the ball to his collapse, is a powerful example of how pain and injury are not proportional to one another. It is only when we process the injury by attaching meaning to it that we express suffering and pain. In this way pain is a form of communication—it allows other people to see what the injury means to us as individuals, and it has survival value in that our expression of pain will produce empathy and assistance.

Beckham sits on the bench with a towel over his head. The injury has not healed, and therefore harm-sensing receptors are still being activated—nociception is ongoing. The information continues to reach the parts of his brain that were activated at the time of the injury, but he is no longer rolling about on the field; the experience of pain has been suppressed to a degree. This is caused by the activation of descending inhibitory pathways that project from the brain down to the spinal cord. The brain can modulate the information coming up from the injured area, and the pain experience is altered. The brain acts like a police officer controlling traffic, deciding how many of the cars coming from Achilles Avenue are to be allowed through. The brain also decides how much attention to pay to the individual drivers of these cars. The need for immediate attention is over for Beckham; the behaviors he exhibited are no longer necessary, nor are they socially or psychologically acceptable. The game must go on and he must retire privately to decide where he goes from this point in his life. This is the loneliness of the pain experience.

The descending inhibitory pathways that project from the brain are poorly understood. Functional magnetic resonance imaging (fMRI), which lights up parts of the brain in response to a painful stimulus, has been used to theorize about which areas of the brain are involved in suppressing pain. Parts of the emotional center of the brain, the hypothalamus, which is responsible for appraising the sensory information, as well as the rostroventral medulla (part of the midbrain), are important for integrating these descending pathways to the spinal cord. Interestingly the same brain chemicals we try to increase with medication when treating depression (serotonin and noradrenaline) are fundamental to the activation and functioning of these pathways. This is why individuals who are depressed often experience increased pain: the lower levels of these chemical neurotransmitters are insufficient to activate the descending inhibitory pathways.

And so we see that not all signals that are being sent from his Achilles tendon are being perceived in Beckham's brain at the point of injury and in the immediate aftermath. Attention to the injury, emotional processing of the experience, expectation, and thinking about the meaning of what has happened—these are all triggering return messages from the brain that regulate and control the information traveling from the injured area and moderate the consequent pain suffered.

I often use the example of Beckham's injured tendon when I explain to medical students the difference between tissue damage and pain. Initially he does not appear to be in pain, although of course we do not know that this is the case and can only surmise from the behavior that he exhibits. He tries to continue to play; the information about the damage has already reached his brain, but until he makes sense of what has gone wrong, we would not recognize his behavior as that of exhibiting pain. This same situation, the mental processing of an injury, is why soldiers injured in battle sometimes do not complain significantly of pain even in the presence of horrific injuries such as amputations. Their immediate concern is to get to safety, and the brain can override the pain experience to facilitate this. It is only when soldiers reach a place of safety that they make an appraisal of their injury. In 1981 President Ronald Reagan was shot in the chest in an attempted assassination. He recalled later that he realized he had been shot only when he felt and saw the blood seeping through his shirt in the limousine on the way to hospital.

These examples of a delayed pain experience can be contrasted with those of people who are caught in a fire. A fire has none of the modulating environment of a battlefield, with its chaos, or the distraction of being hastened to safety by Secret Service agents. Nor, crucially, does the person about to be burned expect the injury. Because the fire is a surprise and the injury unanticipated, the victim deems it catastrophic and experiences immediate and severe pain. Burn injuries are among the most painful given the degree of damage that is done and the widespread activation of harm-sensing receptors on nerve endings.

We all appreciate and understand the kind of pain I have described, caused when tissues are broken, torn, or damaged in some way; it is no different, perhaps, from the experience of primitive or ancient humans, who understood the pain inflicted by an arrow but failed to appreciate the pain from degenerative joints or infected, inflamed tissues. But as we have seen, injury and pain are not proportional, and the outcome of pain is dependent on more than just the degree of damage sustained. This can be harder to understand.

Pain as an experience is influenced by beliefs and expectations as well as psychological factors such as mood and resilience. An individual's culture, genetics, and innate ability to cope with adversity will impact their experience of pain. The rupture of an Achilles tendon in a weekend squash player has very different implications from the same injury in a professional soccer player who has aspirations of captaining a national team in a World Cup tournament; a soldier on the battlefield responds very differently to discomfort than a member of the public who is injured at home, yet they both have the same internal chemistry.

The complexity of the psychological experience of pain is best understood by looking at an individual's reaction when they sustain an injury: their facial expressions, language, and the sounds they make. But this needs to be considered alongside the physical processes taking place; the danger is that we might begin to consider the psychological experience as being separate from the pain pathways I have described. I talked about activation, for example, of the fight-or-flight system. We know that the release of adrenaline as a response to stress can cause an increase in muscle tension and that the increase in muscle tension reduces blood flow (and therefore oxygen supply) to the muscle, which then releases bradykinin in response to being starved of oxygen; the bradykinin then activates more harm-sensing receptors. Anxiety and stress, therefore, which we would regard as psychological constructs, influence the pain experience on a biological basis. If a doctor is kind to you when you are in pain, that pain may be relieved to a degree even before the doctor has given you any pain medicine, because kindness activates the descending inhibitory pathways. Similarly, while hostile social situations or hazardous environmental factors can cause the suppression of pain, as they do for soldiers on a battlefield, in other environments these factors can actually aggravate the pain experience because, as we saw with adrenaline, stress influences the production of hormones, which activate parts of the brain that can either exacerbate or relieve the pain. The differing effects of psychosocial factors on an individual's pain experience depend on how they are interpreted. There is an integral relationship between psychological, social, and environmental factors and tissue injury. This is why, when considering the experience of pain, we use a biopsychosocial model, which proposes that psychological and behavioral processes are mediated by the individual's biology, rather than representing some esoteric force that descends from the ether. In other words, eventually it's all about chemistry.

As human beings we struggle to understand how our behaviors and thoughts influence our biology via hormones and neurotransmitter chemicals. This is why individuals with mental health disorders are so marginalized and poorly understood and are therefore discriminated against, while we readily accept high blood pressure as a disease because it has a physiological explanation, even if most of us would never be able to outline how blood pressure functions within our bodies. Very few individuals understand that normal blood pressure is the product of the output of the heart multiplied by resistance in the blood vessels and that abnormal blood pressure is a result of dysfunction of these blood vessels. We would not respond to somebody with high blood pressure by telling them to calm down or relax their blood vessels. Unfortunately, patients who present with pain that we consider to be in excess of what we would expect, or who have lowered mood, are often told to just get on with it and stop being histrionic.

———————————

Depression and anxiety play a role in the experience and report of pain. Catastrophizing is a psychological construct that consists of excessive rumination, magnification, and helplessness in the face of adversity and involves extreme negative thoughts about one's plight. A higher level of catastrophizing, and consequently feeling a lower sense of control, is an important predictor that a patient will experience severe acute pain or develop chronic pain, when pain persists beyond what would be medically expected. Catastrophizing may be thought of as an exaggerated negative pattern of thinking during actual or anticipated painful experiences. Where patients catastrophize to a high degree they experience greater levels of pain and emotional distress. This abnormal pattern of thinking can have a significant biological impact, as it can increase pain signals and may result in increased sensitivity of the pain alarm, potentially leading to its permanent activation even after the once-damaged tissues have healed and resulting in progression from acute to chronic pain.

Chronic pain can in part be due to the development of pain-related fear, when patients adopt an approach to managing persistent pain by trying to avoid activities that might cause an increase in pain. Patients who have undergone surgery or experienced trauma and who have high levels of catastrophizing, for example, may find it difficult to rehabilitate because of this psychological construct. Their way of coping with pain is to avoid movement; this may have consequences such as the development of deep vein thrombosis when they remain in bed and the blood in their legs pools like stagnant water, as well as lung infections when they fail to cough or breathe deeply following surgery. These patients are also more likely to use higher doses of pain-relief medication.

Preoperative anxiety is one of the most consistent and predictable risk factors for the severity of postoperative pain. One of the key messages that I communicate to trainee doctors who assess patients prior to surgery is to never tell anybody that they will have no pain following surgery. I tell them to explain to the patient that if zero equals no pain and 10 is the worst pain imaginable, then at best we can probably get their pain down to 4 following the operation. If patients have this 4/10 construct in their heads when they wake up and assess the sensory information reaching their brain, it may allow them to cope better with the postoperative pain. If you tell them they will feel *no* pain when they wake up, then any unpleasant sensory information will be appraised in a catastrophic manner.

The importance of this approach is often emphasized to me when I look after patients who have had a laparotomy, which involves cutting through all of the abdominal muscles and tissues to get to the bowels. The incision is often four to six inches long and can extend from above the pubic bone to just below the chest. We use epidural pain relief for these operations; a catheter is threaded via a hollow needle into the epidural space, which lies just above the spinal cord, and the infused local anesthetic bathes the nerves as they leave the spine. This technique usually provides excellent pain relief for what is a massive traumatic incision. These operations also require patients to have a plastic breathing tube in their throat for a prolonged period and also to have a plastic drip put into one of their neck veins. I am always amazed when patients complain bitterly of the tube in their neck and a sore throat, and are almost unmanageable as a consequence, but don't mind the six-inch laparotomy incision. The anesthesiologist's preoccupation with the laparotomy wounds and the epidural often causes us to forget that patients will attach meaning to *any* unpleasant sensory information following surgery, even for issues that we would regard as minor compared with the pain they would be experiencing if they had not had the epidural. This is why patient-centered care is much more successful than clinician-centered care in helping people.

That patients would expect to not feel any abnormal sensations after surgery confuses doctors and nurses who work in the healthcare setting on a daily basis, as people who have been trained in the use of analgesics appreciate quite readily that these medications cannot abolish pain completely. I think it sometimes doesn't occur to healthcare workers that while we have become rooted in a culture that understands a vast amount about how the human body functions, our understanding is sometimes imperfect. When I teach neurosurgeons, for example, I point out that when they look at their hands what they see are structures and nerves that provide sensation to those structures. They understand how tendons located in the forearm make fingers move. It is therefore not surprising to them that if a nerve is damaged at the elbow, the function of the hand is affected. Most laypeople, however, have a more rudimentary understanding of how their body functions, with beliefs that are often fixed in urban legend rather than in science, in the same way that the gasoline engine or the inner workings of a computer are a mystery to me. It is therefore our privilege and responsibility as healthcare professionals to translate medical information into language that paints a lucid picture. I sometimes think that, just as some universities have a professor of the public understanding of philosophy and science, there should exist a faculty for training individuals who aim to bridge the chasm between the science of medicine and its intended beneficiaries. Regrettably, many believe that doctors naturally bridge this chasm.

While patients may not need to understand the details of the circuits and molecules involved in producing the experience of pain, as I have described in this chapter, it is important that they understand how powerfully their interpretation of what they are feeling impacts on their pain. An understanding of the role of thoughts and feelings will enable the person suffering to manage their expectations regarding pain and its treatment. While the distress associated with the pain experience is normal and often useful, as it can provide crucial information and lead an individual to seek appropriate help, a good understanding of what influences pain and why can be hugely beneficial in both managing pain and even decreasing it (by reducing feelings of being overwhelmed and distressed). Pain is, after all, both a sensory and an emotional experience, with the initial information regarding the damage activating various parts of the brain and triggering a complex physiological and psychological response. It is only by addressing the complete experience of pain that we can hope to manage pain and alleviate suffering.

CHAPTER TWO

A Brief History of Pain

The historical record indicates that pain has always been a part of human existence and any organism that has consciousness will be subject to it. The experience of pain is thus primal and facilitates avoidance and education about situations that could lead to death. Inscriptions on parchment scrolls from ancient Greece, Babylonian clay tablets, and papyri from the days when the pharaohs ruled Egypt testify to the fact that human beings have long been preoccupied with understanding and extinguishing pain.

Historically, in the absence of an understanding of the physiology and mechanism of pain perception, we resorted, as humans always do, to creating meaning and learning from experience. Primitive humans probably discovered that if they rubbed the injured part of their body it reduced their pain, and if the injury was exposed to an alternative stimulus, such as cold water or the heat of the sun, pain could be alleviated. Today we take for granted a basic understanding among laypeople of human anatomy and physiology, but to primitive humans pain from internal disease must have seemed quite disturbing. As a result, they formulated narratives and explanations to explain the experience of pain, in the same way that, in the absence of scientific language, creation myths sought to explain the origins of humans and their purpose on Earth.

In primitive societies the intrusion of evil spirits, demons, or magic fluids was often regarded as the reason people experienced pain; the aversive nature of the pain experience coupled with a lack of diagnostic tests to identify its cause is perhaps the reason pain was regarded as evil and demonic. Pain management in these early days therefore involved warding off evil spirits using tattoos, amulets, and talismans derived from animals.

Primitive humans also experimented with various herbs and plant foods that serendipitously were found to relieve pain. It is thought, for example, that intestinal parasites were treated by prehistoric alpine Northern Europeans with a birch polypore fungus. (The fungus was discovered on a 5,300-year-old mummy in the Italian Alps who was found to have intestinal parasites that the fungus actively attacks.) At the more extreme end it has been speculated that Peruvians used trepanning (making a hole in the skull) as early as 400 BCE to treat headaches or epilepsy; the spirit causing the pain exited through the hole. This has been identified as the first surgical procedure, making neurosurgery the most ancient of surgical practices and, as I like to remind the neurosurgeons I encounter, makes them truly prehistoric in their dispositions. Aborigines in central and southern Australia in the late nineteenth century set broken bones using clay and covered wounds with animal fat and bound them with bark or animal skin. We still exhibit elements of this primitive behavior today, when we employ traditional treatments in the form of warming rubs, cod liver oil, and home remedies to relieve pain. Their efficacy has no empirical basis but resides in the collective cultural memory.

When we reach the end of our homespun therapies, we often resort to someone who represents comfort—in my case usually my mother or a wise aunt, even though my dad is a GP. The first healers, according to anthropologists, were women, who were represented by the large-bosomed Gaia, or Great Earth Mother, and acted as both priestesses and sorceresses. Given that women were the conduits of life in the form of the birth process, they were seen as being more connected to human suffering and therefore better equipped to help alleviate it. As in most cultures and throughout recorded time, whenever women originate an industry, men will take control if they feel that power or money can be made from it, and over time the medicine man took over from the medicine woman as the magician responsible for driving out demons. At this stage in history, "medicine" referred less to the administering of physical substances and more to spiritual healing.

In ancient Egypt (around 4000 BCE), painful conditions that stemmed from something other than a physical wound were deemed to be caused by the influence of gods or by spirits of the dead that entered the person through their nostrils or ears while they were sleeping. These demons could be extruded by vomiting, sneezing, or sweating. Even today, despite our advanced understanding of viruses, we are still told not to sit on cold floors or we will "catch cold" and that the source of an upper respiratory tract infection is due to walking outside without a coat rather than the transmission of a virus through secretions. While we don't call the resultant infection a demonic possession, our language nevertheless does not incorporate a modern understanding of why we are afflicted, and the causal relationship is therefore more imagined than real.

The Egyptians developed the first neurophysiological explanation for pain, describing a fluid network that ran throughout the body carrying the essence of life and sensations. They understood that damming the River Nile resulted in poor crops and applied this idea to people; for example, berries of the castor oil tree (a laxative) were used to facilitate unblocking when people were unwell or in pain. The Egyptians had an extensive pharmacopoeia, including the use of honey as a medicine; despite knowing nothing of its specific antibacterial properties, they learned that it could be used to heal infected wounds. Surgery and dentistry were well developed in ancient Egypt. Circumcision of male children was the norm because of the belief that it was cleaner to be circumcised.

The Babylonians some two thousand years later worked on a similar system that saw religion and early medicine run hand in hand. The name Babylon means "Gateway of the Gods," and the gods were believed to decide all matters dealing with human fate, including those of health and sickness. Demons were the invisible bearers of disease, making banishing afflictions a religious affair. That is why Babylonians often used incantations, akin to prayers, in their treatments to relieve pain. The role of priest and physician was therefore rolled into one, "It's the will of the gods" becoming a useful explanation for pain.

The work of these early healers often involved changing their physical attributes (think white coat and stethoscope) as well as their place of residence (think hospitals) in order to exorcize demons and ritualize or formalize the experience; an illness thought to have a spiritual cause could be cured only in a spiritual place. Belief in a therapy, and the subsequent creation of a profession, is often aided by a specific location for the work of the professional, causing a powerful placebo effect by setting the process apart from daily life, endowing it with a certain status, and embedding it within the beliefs of the society and the individual. The advent of agrarian, settled societies resulted in ever more elaborate hospitals–religious edifices. Healing and religion became more complex as more people were needed to run these places, with the inevitable production of hierarchies.

There may have been religious underpinnings to the ancient Babylonian medical system, but it was a profession nonetheless, and was policed accordingly. The fee structure for medical services was divided into a public (for slaves) and private (for nonslaves) system, with a reduced fee for slaves. According to King Hammurabi's code for medical practitioners, if a slave died during surgery, the doctor had to provide a replacement slave; if surgery killed a private patient, the doctor's hands were cut off.

The ancient Chinese focused on the imbalance between yin and yang as the cause for illness and pain, mentioned in records as early as 771 BCE but the concepts of which can be traced much further back, to the Shang dynasty around 1100 BCE. Traditional concepts such as yin and yang have been integrated into modern medical thinking in the People's Republic of China, a process that began in the 1950s. In a typical person yin, which is a feminine and passive force (also regarded as cold and damp and representing the lower body), is balanced by yang, which is the male, positive, active force (upper part of the body and back, warm and dry). The vital energy called qi circulates to all parts of a network consisting of fourteen channels or "meridians," which are connected to the internal organs and their functions. A deficiency or an excess in the circulation results in an imbalance between yin and yang, with consequent illness. If there is free flow of qi, there is no pain; if there is no free flow, there is pain. Acupuncture, in which needles are placed at various points along these meridians, corrects the imbalance by redirecting the flow of qi and therefore eliminates pain and disease. Importantly there is no differentiation when it comes to treating psychological or physical disease in the Chinese system, as both are due to an imbalance in qi.

As well as acupuncture, the ancient Chinese had a pharmacopoeia of over 100,000 substances, ranging from animal matter to human body parts. Research into traditional Chinese medicines, many of which are still used today, is needed to define the mechanisms of action and avoid the toxicity associated with some of the treatments. The use of substances derived from endangered species has been criticized. Pharmaceutical research has yielded few useful therapies; however, proponents of traditional Chinese medicine have indicated that complex interactions between medicines have been passed over, accounting for the lack of effectiveness.

In India, Ayurvedic medicine can be traced back as far as 800 BCE. The word "Ayurveda" translates as "the science of life" or "longevity." Within this practice concepts of disease are related to imbalances in the three *doshas*, or biological energies (*vata*, *pitta*, and *kapha*, or space and air, fire and water, and earth and water, respectively); for example, pain is the characteristic feature of deranged *vata*. The Buddha attributed the universality of pain to the fulfillment (or lack thereof) of desires and remarked that any absence or separation of pleasure resulted in pain, a theory that was later echoed in Greek thought by the philosopher Plato. In the absence of an understanding of neurophysiology and in a society seeking meaning through religious belief, the emotional component of pain was given much more attention than its sensory component or physical cause. The Buddha is often quoted as having said that pain is inevitable but suffering is optional. In recent times, we have all but reversed this prioritization. There is an increasing realization that our pursuit of a material, scientific paradigm has left us turning to nihilism or acceptance of inflexible authoritarian structures. Our medical science is not always as helpful as we would like to believe, and when it fails us we no longer have the emotionally comforting meaning-making mythologies of ancient China, India, or Egypt to fall back on.

Our English word "pain" similarly has a combined religious and sociological etymology. It comes from the Greek word for "penalty," *poine*, and the Latin word *poena*, meaning "punishment." Poena was also the name given to the Roman spirit of vengeance and the attendant of Nemesis, the goddess of divine retribution. For both the Romans and the Greeks the experience of pain was laden with an emotional construct. The Greeks in particular understood the sensory and emotional nature of the pain experience, and in the absence of a sound knowledge of neurophysiology they emphasized the emotional aspects of this experience, much like the ancient Egyptians and Chinese, but coupled with what might be considered an early scientific approach. Plato, in the fifth century BCE, saw pain as both physical and emotional. He believed that human beings experience sensation as a result of atoms moving from the veins to the heart and the liver, but that pain is caused not just by stimulation from a peripheral source; it is also an emotional experience in the soul, which was believed at the time to reside in the heart. Around half a century later, the Greek physician Hippocrates, often called the father of

Western medicine, said, "Divine is the work to subdue pain," but he also sought to find physical rather than divine origins for pain and disease.

It was the Greeks who first proposed that the brain was the center for sensation and reason. This view, however, did not gain widespread acceptance, particularly since the influential philosopher Aristotle, like Plato, regarded the heart as the center of all sensation; the capacity for pleasure was in the blood of the heart, which was also the location of the soul. Pain was similarly regarded as a quality or passion of the soul, and the experience was considered to be the opposite of pleasure and the epitome of unpleasantness. The ideas of Aristotle still hold sway in the world today: we talk about a broken heart rather than a broken brain when it comes to emotional distress and pain, and when we are in pain or emotionally upset what we feel is a racing heart rather than a racing brain. The heart is a singularly impressive organ, while the brain looks like a gray gelatinous mass. It is therefore quite easy to see how the narrative of pain and sensation in general became situated within the beating heart.

The Greeks developed a theory of disease that involved bodily imbalances akin to the Chinese system of yin and yang and the Ayurvedic system of *doshas*. In the case of the Greeks it was four humors: blood, phlegm, yellow bile, and black bile. These substances being in excess or deficient was the cause of illness and pain. Diseases were treated by attempting to restore balance (*eucrasia*) between the humors, often by purging, using emetics, or bloodletting. This idea of the bodily humors proved enduring and was included in the Unani school of medicine in Persian and Arabic societies and persisted in various forms across Europe up until the sixteenth century, when it was challenged by the Flemish physician Andreas Vesalius.

The Greeks and Romans made inroads into the burgeoning fields of physiology and medicine. In the first century CE, the Roman writer Aulus Cornelius Celsus, known for his work *De Medicina* (*On Medicine*), related pain to the experience of inflammation, highlighting redness, swelling, and heat as important components of this process. A contempo-

rary of his, Aelius Galenus (known today as Galen), who was educated in
Greece and Alexandria and who settled in Rome, was the court physician
to the emperor Marcus Aurelius. Galen carried out extensive experiments
on dogs by tying off various bits of their nervous system and observing the
consequent disability. Galen divided the body into a complex network of
sensations that involved soft nerves, hard nerves, and pain sensation. Pain
sensation was regarded as the lowest form of sensibility. Despite Galen's
contribution to delineating the nervous system, Aristotle's concept of pain
as passion of the soul would prevail for twenty-three centuries.

The idea that the heart is the center of sensation was still widely
accepted by the time of the Renaissance but had begun to be questioned
because of the discrepancies between the teachings of Aristotle and
Galen. The fall of the Byzantine Empire led to the influx into Western
Europe of Greek texts and their interpretation by Islamic scholars. The
conclusion that the heart is merely a pump came about as a consequence of the extensive dissection of cadavers (prohibited to Galen by
Roman law, although the dissection of animals was permitted) and
enabled the brain to be reassessed as the seat of sensation.

In the seventeenth century René Descartes, an adherent of Galen's
thoughts on physiology, considered the brain and not the heart to be
the center of sensation. Descartes envisioned a single tube running
from the foot to the brain, which when stimulated by a large fire produced a large pain and when stimulated by a small fire produced a
small pain. This was a radical departure from the humors, *doshas*, and
Aristotelian idea of the heart as the *sensorium commune*, the seat of all
sensation. This Cartesian philosophy, which divorced sensation from
emotion, has produced no end of problems; even today many healthcare professionals and patients mistakenly believe pain is proportional
to the degree of tissue injury. Furthermore, the modern understanding
that pain is influenced by psychosocial factors and by the descending
and ascending facilitatory and inhibitory pathways is still disputed by
many doctors. Many times I have had to explain to a spinal surgeon
why it is that one patient's complaint of pain differs from another

patient's after the same operation for the same disease. It is extraordinary how a four-hundred-year-old idea can persist in the face of overwhelming evidence.

Aristotle described nerves in his book *De Anima* (*On the Soul*) as being controlled by and originating in the heart. The word for "nerve" in Greek means "sinew," and it is easy to see how a nerve and structures such as ligaments and tendons, which link to muscles, can be mistaken for the same thing (a mistake I often made as a medical student during exams when we had to identify these structures in cadavers). Galen, who criticized opinions on anatomy that were derived without dissection, recognized the spinal cord as an extension of the brain and documented that nerves control sensation and movement. He thought of nerves as hollow tubes through which the body's vitality (*anima spiritus*) traveled. "White, soft, pliant and difficult to tear" was the description of nerves put forward by Avicenna (also known as Ibn Sina), a Persian physician in the early eleventh century. Alessandro Benedetti, surgeon general of the Venetian army in 1497, described the senses as being distributed by means of the nerves, similar to the way the roots of a tree distribute water and nutrients.

Dissection of nerves revealed that they did not have a cavity, as Descartes suggested, but there was no satisfactory alternative explanation for how they worked until the late eighteenth century, when connections were made between experiments with electricity and the function of nerves. Luigi Galvani—a philosopher, physician, and physicist—described the role of electricity in the nerves of dissected frogs in 1791, and Henri Milne-Edwards (1800–1885), a French physician and zoologist, conceptualized the body as a factory with different physiological systems having different functions; the language of science often takes on the zeitgeist of the time. The development and refining of the microscope (first invented in 1590) and the use of silver chromate to stain nervous tissue in 1873 aided the understanding of the structure and function of nerves. During the nineteenth century, physiology emerged as an experimental science with the publication in 1838 of the concept that all living things are composed of cells and the idea of homeostasis (the maintenance of the internal balance of physiological processes such as temperature regulation and breathing).

The development of physiology as a scientific discipline resulted in theories about pain transmission and perception based on neurophysiological studies. The specificity theory (1859), in which pain has its own sensing apparatus (which is why it is called *specific* rather than *intensive*), had been posited as early as the second century by Galen and picked up throughout the centuries, notably by Avicenna and Descartes. In the nineteenth century this theory was tested by selectively destroying parts of the nervous system in order to see which sensibilities were affected. This was the foundation of neuroscience, although, with tools such as electron microscopy and functional neuroimaging, modern neuroscience embraces a more complex approach to molecular biology, computational neuroscience, and electrophysiology. One can, for example, label and track specific red blood cells, imaging in real time the different connections made in individuals who have chronic pain compared to those without when subjected to a standard stimulus or distraction techniques, thereby outlining the specific neural circuits that are altered in pain states.

The intensive or summation theory (1895), which is Aristotelian in origin, asserts that pain is the result of excessive stimulation of the sense of touch. By the end of the nineteenth century, there was conflict between the *specific* and *intensive* theories because they propose different sensing mechanisms for pain (a specific apparatus versus an overstimulation of touch), both of which move away from and deemphasize the traditional Aristotelian concept that pain is an emotional experience. In 1906 the neurophysiologist Charles Sherrington discovered the receptor present on free nerve endings; this is the nociceptor that transduces the injury into an electrical signal. With that, Sherrington laid the argument to rest: specificity won.

The emotional component of the pain experience was supported and espoused by philosophers and psychologists well into the nineteenth century, but not by scientists. It is interesting to note that as we become scientifically more advanced and ostensibly rational, we tend to separate the emotional from that which we can measure and scientifically study. This continues today in terms of clinical practice, where the emotional component of the experience of pain is neglected in preference to being able to study receptors and pathways and abnormal structures. In clinical practice there is still an uneasy relationship between the work of pain management psychologists and that of physicians.

Theories of the system behind the experience of pain continued to abound as the twentieth century progressed. The pattern theory of pain espoused in 1929 suggests that pain is produced by intense stimulation of nonspecific receptors rather than a specific nerve for pain. The central summation theory of 1949 is an offshoot of the older summation theory and states that intensive stimulation from nerve damage activates nerves in the spinal cord that reverberate, triggering the periphery nerves and causing a closed loop. The "fourth theory of pain," introduced by James Hardy, Harold Wolff, and Helen Goodell in the 1940s, sought to reintroduce psychological factors within the specificity theory. They saw pain as a combination of perception and reaction—a complex physiopsychological response that involves cognition as well as past experience, culture, and psychological factors. They used this information to explain why people have different pain thresholds and why Descartes's model of pain commensurate to injury is inadequate. The sensory interaction theory (1959) proposes that there are two systems involved in the transmission of pain and other sensory information: a slow system and a fast system. Interaction between the different fibers was thought to be the reason pain is experienced. This theory correctly identified the different nerve fiber elements of the pain alarm, but not how it works.

It was first recognized during the Second World War that individuals who are injured in a military setting often experience much less severe pain than those injured in a civilian setting. Doctors therefore realized that there must be modulation of the sensory information that is transmitted to the brain at the level of the spinal cord, and that this modulation must be initiated by the brain to produce different pain experiences. This would grow into the gate control theory that Ronald Melzack and Patrick Wall proposed in their paper "Pain Mechanisms: A New Theory," published in *Science* in 1965. The theory is not recognized by most patients and doctors today, who instead still hold on to Cartesian theories of pain that are four hundred years old.

In the 1950s Melzack and Wall proposed that the specificity theory was supported by evidence of a specialized apparatus within the nervous system for the experience of pain, but there was not a direct pathway from periphery to brain—it was not the hollow tube proposed by Descartes nor a simple uninterrupted electrical cable. They suggested that there must be something at the level of the spinal cord itself that influences neural activity. They therefore proposed in 1965 that there is physiological specialization of the apparatus communicating harmful information and that this information is gathered, made sense of, and changed at the level of the spinal cord, and is also influenced by descending pathways from the brain, which are the psychological factors important in pain. This is now commonly referred to as the gate control theory, whereby an electrical impulse triggered by some form of harm can be modulated at the level of the spinal cord by a nonharmful stimulus. So, for example, if you bump your leg, triggering a pain impulse, rubbing your leg will cause the activation of another nerve fiber, which competes with the initial pain signal for access to the brain.

In Western industrialized societies there is a tendency to divorce the emotional component of the pain experience from the physical, reducing pain to a dualistic Cartesian view where the mind and the body are separate, where we exist as an animal with physical processes, and a psychological being exists outside of us, ethereal and esoteric. This is why we have difficulty understanding pain in the absence of tissue damage and why we also struggle to understand diseases such as depression and psychosis, which are due to dysfunctional brain neurotransmitters but which our current medical technology (MRI scans, blood tests, electrocardiograms, etc.) cannot definitively measure. It is therefore also difficult for people to understand why a small injury can produce significant pain; in a Cartesian world, where mind and body are separate, a small injury should produce a small pain and a large injury should produce a proportionally greater amount of pain. We fail to appreciate the emotional aspect of the pain experience and the ability of the brain to amplify or diminish the signals initially produced by the injured tissue.

For most of humanity's history, pain has been considered an inevitable consequence of existence, and it was therefore the meaning of the pain, rather than whether or not you had pain, that mattered. In the increasingly scientific and industrialized Western society of the nineteenth century, this view shifted; individual pleasure and pain became increasingly important, bringing about a change in the way pain was viewed. Despite advances in the understanding of how the human body worked and, crucially, how it could be repaired when problems arose, the experience of pain was still a significant barrier to performing more intricate surgery. With the demonstration of anesthesia for a dental procedure in October 1846 by William T. G. Morton it became possible to reduce the experience of pain and therefore perform more invasive operations. James Young Simpson's use of chloroform during childbirth in 1848 revolutionized obstetric care.

It may seem strange to us today that there was a lot of debate in Europe with regard to whether or not it was ethical to operate on an unconscious patient, but we must bear in mind that the state of unconsciousness induced under anesthesia was not well understood at the time, nor was it known what long-term effects it might have on the individual's mental process. While there were once questions as to whether we even *should* treat pain, in modern times we have moved dramatically toward aggressively managing pain. The Royal College of Anaesthetists still takes the words of Hippocrates as their motto—

"It is divine to alleviate pain"—but it is possible that, while our scientific techniques and processes have moved on, we have in fact lost sight of the bigger pictures that Hippocrates himself understood two millennia ago: that alleviating pain involves more than simply treating the body.

CHAPTER THREE

Give Me Something for the Pain . . .

Because pain has been our constant companion throughout history, relief from pain and suffering has also been relentlessly chased and our treatments for pain have always been influenced by our politics, culture, available technology, scientific understanding, and the meaning attached to pain.

Every autumn for the past five years, at the beginning of October, I have made the journey to the central campus of the University of Manchester to lecture the second-year medical students on pain management. This is no small task. My lecture takes the students through the management of acute and chronic pain, covers the action of local anesthetics and other analgesics using examples of the different classes of drugs, and demonstrates the key components of the biopsychosocial model of pain. I have sixty minutes to cover the breadth of my specialty, including an introduction that serves to outline where pain medicine fits into the pantheon of medical specialties.

Pain medicine sits uncomfortably amid the subspecialties of anesthesiology. Most other anesthesiologists can't quite understand our choice to be pain physicians (to see distressed patients and sit in clinics) and question our motives (the desire for lucrative private practice, an inability to focus, too much imagination). Sometimes, such as when inserting difficult-to-place epidurals and when offering advice on complex pain patients in the preoperative setting, we are perceived as helpful. But we are also viewed with suspicion and curiosity because we choose to work with a group of patients that most anesthesiologists try to avoid. Compared to anesthesiologists who spend all their time in the operating room, we also have a greater appreciation for running a practice and more familiarity with patient complaints. So we cause anxiety at department clinical governance meetings by challenging the sometimes narrow viewpoint of our colleagues.

Pain medicine is the only subspecialty of anesthesiology in which outpatient clinics are the norm. The skill set required to work in pain medicine therefore departs radically from that of the mainstream discipline of anesthesiology, and includes the ability to actively listen and withstand the emotional onslaught of prolonged engagement with highly distressed individuals. Doctors working in pain clinics need to be able to empathize with patients and manage conflict, disappointment, and expectations effectively. And this extends beyond patient relationships; a life spent working in pain medicine involves nurturing the capacity to work in a collaborative fashion with allied healthcare professionals—physiotherapists and psychologists—as well as developing meaningful relationships with general practitioners in the community.

Many doctors, at great cost to both themselves and their patients, see pain as merely a symptom, a means to formulate a biomedical diagnosis rather than an end or a focus in itself. I have repeatedly emphasized that pain is an experience; it is an experience of feeling unpleasant sensations from our body, which are interpreted by multiple areas of the brain, areas that are responsible not just for registering sensation but for the way we think and feel. Processing abnormal feelings, either physical or emotional, makes us distressed, anxious, and depressed. The pain of a broken heart is therefore no less distressing than that of a broken leg. The way we think about pain, the medicalization of it and whether doctors believe that it should be treated or not, influences how we behave toward those abnormal sensations and the action we take, or do not take, to alleviate them.

As our scientific understanding and technology progress, we perform more and more complex and invasive surgeries, and we demand and require the patient to cooperate psychologically with these insults. People generally don't have the ability to hypnotize themselves and inhibit the sensations when a surgeon slices through their sternum in order to perform open-heart surgery. (The use of hypnosis to perform a chest incision has been described in scientific publications.) Nor do they have the ability afterward to rationally understand that, since the aim was to help pump blood around their heart, the sensations coming from their sternum that has been sliced open are perfectly acceptable. We therefore must resort to medications that enable patients to undergo these procedures. Anesthetic agents that come in gas or liquid form, such as propofol, make you completely unaware of what is happening. We can use medications such as local anesthetics to block sensations from reaching your brain, both for the surgical procedure and for pain management postoperatively. We use opiate medications postoperatively because they affect the way you interpret the sensations from your body: they make you care less. Opiates have been called the perfect "whatever" medication because they allow you to ignore the messages that are coming from your body.

The biological purpose of pain is to alert the organism that harm has occurred, because if the damage is ignored, the open wound will become a conduit for infection, which can lead to death. It could be argued therefore that the purpose of pain is somewhat superfluous in the context of controlled trauma, such as an operation. Consenting to elective surgery is essentially consenting to having a surgeon cut through your skin and tissues and bones. The primitive nature of the pain alarm, however, means that regardless of how cognitively aware you are of why you have sustained trauma, the alarm still goes off. When treating pain we therefore try to tackle the different processes of the pain experience, as I explained in chapter 1: where the tissue is damaged (transduction), the network sending the signals of pain to the brain (transmission), the brain itself (perception), and the pathways that can put the brakes on excitable pain nerves (modulation). There are a number of ways that we can do this, but the most common, and most frequently requested, is by prescribing various drugs.

Before talking about drugs, however, it is worth mentioning that there are psychological techniques in the preoperative setting that can be used to help patients manage their pain. Knee schools have arisen to facilitate improved pain relief and better outcomes from knee replacement surgery; these prehabilitation schools teach patients that even though their knee has been successfully replaced, they will still have some pain after the surgery, but the pain is not an indication that the surgery has been unsuccessful. This facilitates rehabilitation by "normalizing" the postoperative abnormal sensations from the knee, and patients have been found to have an increased ability to mobilize postoperatively when furnished with this information. Patients sometimes compete with one another to see who can mobilize quicker. No magical pharmacological therapy has been administered; it is merely the change in perception and the management of expectations that facilitate the individual's recovery.

When I talk to medical students about using medication to manage pain before, during, and after surgery (referred to as the acute perioperative setting) or in the management of chronic pain, I emphasize that there is no drug in our current armory that will completely abolish the experience of pain—apart from local anesthetics that can *temporarily* block all transmission of electrical signals. Local anesthetics are not suitable for long-term use, however; they are limited in their duration of action when administered as a single injection and require ongoing infusions through plastic tubes placed near nerves to facilitate prolonged use. These infusion devices present a conduit for infection and therefore are limited in how long they can be used. Beyond anesthetics, other pain-relieving medication will not stop all the electrical signals to the brain triggered by chemical, mechanical, or thermal damage. Regardless of how much or how little information is transmitted, the information that does get through will be perceived and interpreted according to the unique expectations, beliefs, mood, and previous experience of the individual. I therefore urge the medical students to always attempt to manage the patient's expectations of the degree of pain relief an analgesic will facilitate.

The same principle applies to the pharmacological management of pain either in the perioperative setting, for those who have had trauma or are about to have elective surgery, or for those who suffer with chronic pain. Patients may have the expectation that following surgery they will have no pain, although they may not say so out loud, and when they do wake up with uncomfortable and abnormal sensations, which they interpret as pain, their immediate perception is that the surgery has not gone according to plan and that something is desperately wrong. Failure to tell patients who are anxious or catastrophize that they will feel these sensations following surgery will lead to a poor perioperative pain experience and can undermine even the best-placed thoracic epidural or judicious use of pain medication. Expectation management around the use of pharmacological agents that do not completely abolish pain following trauma or surgery is therefore fundamental.

―――――――――

The anesthesiologist is the forgotten person in the surgical drama, quietly assessing patients' risk of having major surgery, optimizing the function of organs preoperatively with fluids and medications, obliterating awareness of the sound of a bone saw, and ensuring patients' comfort following the operation. But most people remember only the surgeon.

In the triad that makes up anesthetic practice, ensuring a lack of awareness or recall of the surgeon slicing through the tissues is one goal. Another, not always required, is muscle relaxation (when cavities such as the abdomen must be entered, requiring the protective muscles to be split). But what is fundamentally essential is the management of pain in the perioperative setting. When somebody is asleep or anesthetized, they cannot experience pain, which requires consciousness; the body, however, will still react to tissue damage. One of these physiological responses to injury is an increase in heart rate and blood pressure, which can lead to an alteration in the patient's state of consciousness. They may even become aware of their surroundings as their level of consciousness changes due to the stimulus, which is a disaster if they are unable to move or communicate what they are feeling. Anesthesiologists therefore administer pain medication intra-operatively in order to reduce this neurological excitability and also to ensure that the person is as comfortable in the postoperative setting as they can be.

For some operations patients can be awake, needing only a local anesthetic to block the transmission of information. The only naturally occurring local anesthetic is cocaine, which was used by early Peruvians to assist them with living at altitude by increasing energy and reducing muscle aches and its euphoric effects were utilized in religious ceremonies. Cocaine and was introduced to the West in the latter half of the 1800s, where it was first used in a medical setting for an eye operation. The synthetic drug procaine (marketed under the trade name Novocain) is related to cocaine but is much less toxic; it was formulated in 1904, shortly followed by lidocaine, which was developed in 1943 and came to market in 1947 as a local anesthetic less likely than procaine to produce hypersensitivity reactions. Lidocaine is still a commonly used local anesthetic agent, along with bupivacaine, another member of the cocaine family.

When operating on a conscious patient, a needle can be inserted near a large nerve, surrounding it with injected local anesthetic, which will provide pain relief for around eight hours or so. The local anesthetic crosses the membrane of the nerve and blocks the sodium channel of the nerve from the inside, which means it cannot fire for as long as the local anesthetic sits there like an unwelcome guest. But the needle can inadvertently puncture the nerve, damaging it permanently and causing neuropathic pain for the rest of the patient's life, although this occurrence is rare.

Ultrasound has become the chief means by which we aim to deposit local anesthetic safely around nerves; by visualizing the tip of the needle, we can prevent it, theoretically, from going in too far and injuring the nerve. Ultrasound images, however, often look like black-and-white impressionist paintings, and so nerves that are supposed to look like a bunch of grapes in cross section are sometimes not clearly seen. Nevertheless anesthesiologists like new toys and are very keen exponents of this technique. They are often ridiculed for their obsession with the latest electronic devices (normally made by a company named after a fruit also implicated in poisoning Snow White) as well as coffee (our department has a coffee machine worthy of a professional barista), and either cycling, triathlons, or running marathons in various cities around the world.

Local anesthetic can be injected either at a peripheral nerve near the tissue damage (as described earlier) or at the nerve's origin in the spinal cord. Many patients who have knee replacements have a spinal anesthetic, a process in which a needle is inserted into the area where the cerebrospinal fluid circulates and through which small amounts of local anesthetic are injected. This effectively numbs all the nerves that provide sensation to the legs, rendering the patient numb from the waist down and therefore able to be awake while the knee is being replaced—although patients might be given some sedation, mainly to avoid the disturbing stream of consciousness that can emanate from some orthopedic surgeons. This technique can probably be safely practiced for operations up to and including a hernia operation done just to the right or left of the belly button, and even for cesarean sections. The problem is that the higher up the body the local anesthetic spreads, the more likely the patient is to stop breathing. There is a fine line between injecting the local anesthetic in a great enough quantity to block the relevant nerves for the area being operated on and injecting so much that the patient suffocates.

The spine is not straight but has a concave curve in the lumbar region just above your bum, then is straighter in your middle spine (called the thoracic spine) before having a convex curve in your neck. The trick is to put the local anesthetic in below the level where the spinal cord ends, usually taken as a point below a line drawn between the tops of your hips. As I'm sure you can imagine, however, this is a variable target depending on your body type, and sticking a needle right into the spinal cord is never a good idea; all the nerves in the cord run very closely together, and with one needle track you can injure the nerves that come from multiple areas of the body.

An alternative to inserting a needle into the fluid-filled space around the spinal cord is to insert it in the space that lies between the last ligament of the spinal bones and the lining of the spinal cord, known as the epidural space. This is what we call a potential space because the two layers are separated by dilated veins and therefore must be accessed via a loss-of-resistance technique. A needle is inserted into the last ligament layer between two adjacent vertebrae and then a syringe filled either with air or saltwater is attached. When the needle gets to the potential space, the syringe plunger suddenly gives way; this is called loss of resistance. Too far and the needle will make a giant hole in the lining surrounding the spinal cord, causing the cerebrospinal fluid to leak and giving the patient a massive headache; not far enough and the catheter won't go into the space, so the local anesthetic won't bathe the nerves as they leave the spinal cord. This technique is relatively easy to do in the bottom region of the spine— unless the purpose is to ease the pain of labor, in which case it's a bit like trying to thread a needle in the dark. For operations where the incision will be higher than the belly button, the anesthesiologist has to insert the epidural higher up, level with the bottom of the shoulder blades. The challenge then is that the two adjacent vertebral bodies are packed tightly on top of one another, virtually obliterating the space the needle needs to enter.

As pain consultants we are sometimes asked by our anesthetic colleagues to use an X-ray machine to guide the epidural needle into these tricky places. Epidurals for pain relief in surgeries such as cutting out a cancer in the food pipe can influence whether you live or die, but for other surgeries a careful assessment is made to establish whether the risks of the epidural, which can include permanent nerve damage, are justified. Sometimes epidural catheters fall out and become infected, resulting in an abscess around the spinal cord that, on rare occasions, can lead to paralysis.

Another anesthetic-based drug that is sometimes used to manage chronic pain and pain in the postoperative setting is ketamine. It is a phencyclidine derivative—basically PCP. It was developed in 1962 and used extensively in the Vietnam War as an anesthetic agent because in the middle of a paddy field you can't intubate the airway to help an injured soldier breathe. Ketamine, while rendering you insensate, does not disturb your ability to protect your breathing passage; there is therefore no risk of your aspirating your stomach contents while unconscious. Ketamine is often used now in trauma because it does not cause a massive drop in blood pressure like other anesthetic agents, such as propofol. It produces what is called dissociative anesthesia, where you are unresponsive to pain but completely aware of your surroundings—essentially you enter a trance-like state. We also use it in children for painful procedures; the drug can cause nightmares, which we try to prevent by giving diazepam concurrently, but the risk of a nightmare is more tolerable than giving children drugs that could stop their breathing. Ketamine is also used to medicate horses and is a recreational drug of abuse.

We use ketamine in the perioperative setting because it blocks receptors at the spinal cord level that are involved in amplifying pain in people who have long-standing pain or when the operation involves such a big incision that it is likely to produce a massive response in the pain alarm system. We give it as an infusion intra-operatively and continue it postoperatively to reduce the amount of opioid medication the patient needs to take. As with most anesthetics, however, it is really a short-term solution. Long-term use of ketamine damages your brain and bladder, turning the latter into a small, irritable, and painful organ.

———————

Local anesthetic blocks are a triumph when the patient is pain-free in the recovery room or, despite having serious heart or lung disease, is able to have complex shoulder surgery without general anesthesia. Unfortunately, the local anesthetic tends to wear off in the middle of the night, so unless it is given through a continuous infusion the patient is left in agony. Postoperative pain management is complex and fraught with difficulties.

The management of perioperative pain went through a revolution in the 1990s, when it became clear that we were failing to implement policies to treat postoperative pain in a consistent way. We realized that untreated pain has significant effects on the outcome of surgery. Being in pain results in a physiological response; for example, it can cause an increase in heart rate, which can cause a heart attack. If you have had a big cut in your abdomen, pain might prevent you from breathing deeply because with every breath the pain alarm becomes louder—and so you take shallow breaths, resulting in the bottoms of your lungs collapsing and filling with fluid, which then becomes colonized with bacteria; you may develop pneumonia and potentially respiratory failure. Similarly, when you are in pain you are unlikely to want to move, but staying bed-bound causes blood to pool in your legs. When this happens clots form and bits of clot can break off and travel like missiles to your heart, blocking the outflow of blood. This is called a pulmonary embolism and can result in death when the embolus is large. In these circumstances the clot will resemble a horse's saddle in the part of the blood vessel that it lodges in. Usually the clot will break off when the patient first gets up to go to the toilet and bears down, often the cause of an instantaneous and undignified end.

As well as altering a patient's behavior, pain activates hormones and chemicals that affect the body's ability to heal. When we are stressed we naturally secrete cortisol, a hormone that breaks down protein to make sugar available for muscles to use as energy, fueling flight from our plight. Untreated pain results in an increase in this hormone, with the effect that it begins to break down muscle, leading to loss of muscle mass, which affects breathing and results in delayed wound healing. This can have catastrophic effects; for example, if a surgeon has joined two segments of your bowel after excising a cancer, a delay in healing and potential breakdown of the join can lead to a leak inside the abdomen, causing inflammation, sepsis, and potentially death.

In the 1990s and 2000s many professional healthcare bodies representing surgeons and anesthesiologists encouraged hospitals to provide specialist acute pain services. Pain was labeled "the fifth vital sign," and aggressive pain management was considered both clinically important and a fundamental human right. (The other four vital signs are temperature, heart rate, blood pressure, and respiratory rate.) In retrospect, however, the zealous pursuit of this goal and linking pain relief to hospital quality outcome measures may have contributed to the opioid epidemic, which I detail in the next chapter.

The hospital I work at has a large acute pain practice consisting of specialist nurses led by a nurse consultant. The pain consultants play a role in managing the more complex cases and are responsible, together with the nurses, for monitoring and implementing policies to make sure that every patient has adequate pain relief and that the latest advances in pain management are implemented in a systematic way. The surgical patient population is growing progressively older, meaning that physiological tolerance to injury is lower and surgeries are more complex, and the more complex the surgery, the more complex the perioperative pain management requirements. Good pain management requires pain-relieving medication protocols that can be consistently prescribed for standard types of operations, while bearing in mind potential side effects.

The main aim in perioperative pain medicine these days is enhanced recovery. We used to get people to stay in bed for ages following surgery, until they felt better, but the philosophy of postoperative care now is to get people up and about as soon as possible. Patients who have had major bowel surgery, for example, are up on the first day, and the aim of their pain management is to provide good pain relief without constipating them up to their eyeballs, which would potentially affect the joining of the bowel segments and suppress the immune system. Getting up and about also reduces the risk of developing chest infections and deep vein thrombosis and promotes wound healing, but this can be accomplished only if the blaring pain alarm system is managed effectively.

Pain assessment is fundamental to being able to treat the experience. The Dutch have a saying: You can't manage what you don't measure. As part of their preceptorship courses, newly qualified nurses are trained by the specialist pain nurses in the assessment of pain in order to understand the phenomenon as a complex experience influenced by the individual's unique psychological and social context. Therefore, the medication management of perioperative pain or pain due to trauma takes a multimodal approach (targeting the different parts of the biological pain alarm system—the nociceptor, spinal cord, and brain) supported by an infrastructure to deliver and audit pain management; this usually comes under the auspices of a specialist pain clinic. The aim is to target the different parts of the pain alarm system in order to modify the information received by the brain.

In a healthcare system that is largely biomedical we resort to medications to try to achieve this aim. The evidence base for modulating pain pathways using education and psychological techniques is less well developed, and while there is very little money in marketing a behavioral therapy, there is a lot to be gained from developing and selling medication. In the 1990s and 2000s, opiates were aggressively developed and marketed; companies generated potent synthetic opioids as well as novel delivery systems (including fentanyl lollipops and nasal sprays). The ubiquitous and sometimes indiscriminate use of opioids led to an undetected epidemic of opioid misuse, which is now coming to light.

As part of a multimodal pain-relief strategy we give patients anti-inflammatory medications. Anti-inflammatories have been used in medicine for millennia; the agent salicin, the precursor to aspirin, is naturally occurring in willow bark and was known as far back as Hippocrates to be effective for the treatment of fevers. The first anti-inflammatory medication was synthesized by the Bayer pharmaceutical company in 1897, but the mechanism of anti-inflammatory drugs was elucidated only in the 1970s, which led to the development of newer agents. We were already marketing anti-inflammatories that had been developed without the necessary degree of understanding, and many of these drugs are still used today. Examples are ibuprofen (the patent for which was filed in 1961) and diclofenac (patented in 1965), which act by blocking prostaglandins from attaching and triggering nociceptors. If trialed today, many of these drugs would probably never get past an ethics committee or be given a license for use, because of their potential for catastrophic complications, including stomach hemorrhage and kidney failure. Even ibuprofen, a common over-the-counter drug available in pharmacies and supermarkets, can cause indigestion, stomach ulcers, and in some individuals heart and kidney failure.

In an attempt to design safer forms of anti-inflammatory medications, drugs were produced that acted only on the specific enzyme involved in producing the inflammatory chemical that triggers nociceptors and therefore produces pain. But we found out that if you take these medications for a long time you have an increased risk of heart attacks and strokes. Anti-inflammatory medications are difficult to give to patients who haven't eaten because blocking the prostaglandins generally stops their protective function in the lining of the stomach, allowing acid to potentially eat through the stomach lining, causing an ulcer with resultant bleeding or a hole in the intestine. These drugs are probably safe to give for only five to seven days after an operation.

If we can't anticipate how much blood loss the patient will suffer in surgery, it becomes unsafe to give anti-inflammatory medication at the beginning of the operation because these medications will further compromise their kidney function. Anti-inflammatory medications also affect the stickiness of platelets, which are necessary for clotting, and therefore pose a risk for increased bleeding. Brain operations, for example, are problematic in terms of prescribing anti-inflammatory painkillers because if the patient does bleed it will be into a fixed box containing the brain and the increased blood will crush the brain tissue.

Orthopedic surgeons believe anti-inflammatories negatively influence how quickly bones heal. The balance of evidence from experiments on animals suggests that prostaglandins favor bone formation. Nonsteroidal anti-inflammatory drugs, or NSAIDs, might therefore be expected to inhibit bone formation because they inhibit prostaglandin formation. Experimental work in animals seems to demonstrate, at best, that the impact of NSAIDs on bone healing after a fracture or operation is uncertain. The evidence is contradictory, however; there are issues over drug dose and duration; and there is little evidence from clinical practice to support extrapolation to humans.

Trying to block the chemicals that are released when the body is damaged is challenging, and the story of medication for pain is beset with our inability to be precise in blocking harm-sensing information. If in the future we could stop these chemicals from attaching to pain receptors and thereby triggering the electrical signal that the brain interprets as pain, we would not experience pain after surgery—but we are not there yet.

Paracetamol, known in the US as acetaminophen, is an oddity as a pain-relieving medication—widely used but, again, poorly understood. Paracetamol is a fever-lowering drug primarily and a very mild painkiller that we think works on cannabinoid receptors (receptors involved in mood, appetite, pain, and memory, which are widely found in the body) or possibly on a mystery enzyme. But the truth is we don't really know how it works. It was accidentally discovered when looking for a cure for intestinal worms. The story goes that two researchers were using naphthalene (a hydrocarbon found in plasterboard, lead batteries, dyes, and rubber) as a treatment for killing gut parasites when a pharmacist accidentally delivered a different chemical to them. When they used this chemical in their experiments, they found that it didn't kill the parasites but did lower fevers. The serendipitous mistake by the pharmacist was uncovered and the researchers produced a drug called Antifebrin (from the word "febrile," as in "fever"). The drug that was formulated lacked purity, however, and individuals experienced unpleasant side effects such as nausea and headaches.

It wasn't until the 1940s that the toxic metabolites were separated from the positive effects of paracetamol. The drug Trigesic, a mixture of aspirin, caffeine, and paracetamol, was first marketed in the US in 1950, but when three patients developed death of all their blood cells it was thought to be related to the paracetamol. In the UK paracetamol was produced as a pure formulation and marketed as Panadol. The first scientific study of Panadol concluded that it was not particularly effective as a painkiller compared with codeine; slowly, however, paracetamol was combined with other medications, increasing its effectiveness as a painkiller and reducing the side effects, and as the use of aspirin declined as an over-the-counter analgesic, paracetamol took its place. Today paracetamol (marketed as "Paracetamol—the GP's choice") remains the first-line agent for many people who suffer with aches, pains, headaches, or fever, despite having only a modest effect on the first three. The story of paracetamol highlights the development of a drug not by first understanding pain or its mechanisms but by observing an effect and then working to establish safe use.

The main class of drug that we use to reduce and confuse the messages of harm sent to the brain are opioids, medications that bind to opioid receptors present in the body. Opiates have been used since the time of the ancient Greeks, mainly to treat diarrhea, insomnia, and pain. There are opioids that are naturally produced in the body (endorphins and enkephalins) that attach to these receptors. Opioids either come from the poppy (and are known as opiates), such as morphine, or are synthetic. We used to think that all we were doing with the opioids was benignly hyperpolarizing the nerves. This involves making the nerves less excitable by reducing the ability of sodium and potassium ions to flow into and out of the cell; the cells are made more negatively charged and therefore cannot fire off an electrical signal—they are too lethargic and depressed. The amount and nature of the transmitted information are thus altered in the spinal cord and consequently by the brain cells receiving the information.

We have always been aware that the reason opioids are abused is that they affect multiple pathways in the brain, producing changes in mood and motivation. Their perceived pain-relieving effect is now thought to be due to their effect on the emotional component and individuals' reaction to pain, blunting the ability to actively manage or think about the sensations they are experiencing, rather than by blocking nerves or reducing inflammation. Opioids affect how you feel about your pain and your life in general, stifling emotional responsiveness. My patients, who have chronic pain and have had the opportunity to reflect on their relationship with opioids, often say that when they take their opioids they still have pain; they just don't care as much about it. There are patients who like this effect, and they may be at risk for addiction. And so the theme of empirically using a substance that has effects we don't fully understand continues. Fundamentally this problem arises because we don't really understand pain to a sufficient degree to design drugs that are specific, and because we are using a medication to treat an experience.

I spend approximately 20 percent of my working life managing people with complex pain problems on the wards—either individuals who are struggling with their pain management following surgery or individuals with chronic pain who have been admitted to the hospital with a flare-up of their pain. I usually start my ward rounds at about eight o'clock in the morning, at which point I receive a sheaf of referral forms—and to say that there is a paucity of detail included about the patient would be generous. The referral, which generally appears to be inspired by the admonition of Polonius to Laertes that "brevity is the soul of wit," usually pithily states, *This patient has had pain for a long time. They have had an operation. Please see with regard to pain management.*

Fortunately, in our hospital we have an extremely robust electronic patient record system, and if the patient has been seen by anyone in the hospital, their outpatient letters will be in the system. One of my first actions is to see whether the person has "form." The outpatient letters, which often do not identify the operation performed, detail in coded language the hidden psychosocial distress associated with a long-standing pain condition and the journey that has led them to have an operation with a very low probability of success. Knowing this about the patient before you even enter the arena is fundamental to establishing a relationship, the aim of which is to get the patient up, moving, understanding their pain, reducing their distress, and reducing their reliance on opiates.

For the average patient the acute severe pain following any surgery usually decreases to manageable levels reasonably quickly. By three weeks following surgery most people need only over-the-counter pain relief. But what about those who don't take this path? The challenge that we face as anesthesiologists and consultants working in pain management is where surgery or trauma occurs in an individual whose pain alarm system is already dysfunctional—the person who has established chronic pain (see chapter 5, "Pain with No Injury"). Particular care needs to be taken with these individuals, as well as those who are having surgery that is so destructive—in terms of the amount of tissue that is damaged and the body cavities that are injured—that it causes a massive disruption in the patient's normal physiology and prolonged hyperactivation of the pain alarm, potentially leading to the development of chronic pain. These patients require an augmented approach, using additional drugs, together with the more usual opiates, local anesthetics, paracetamol, and anti-inflammatories.

The drugs that we employ in the treatment of complex persistent pain conditions aim to modify the dysfunctional pain alarm system—turning it down, muffling it without suffocating the patient in the process. Where there is crossover between dysfunction of the pain alarm system and other conditions we may also borrow drugs from other disciplines that attempt to make nerves behave or change brain chemicals. But our understanding of the molecular mechanisms of pain is limited, and the medications we borrow from other disciplines have simply been serendipitously discovered to distract the nervous system rather than being focused on a specific target.

One example of this is the use of antidepressants to manage chronic pain. The original drug used in this regard is amitriptyline, which is one of the older generations of antidepressants and is no longer in common use because of its high risk of overdose. When I worked in South Africa at a hospital in one of the townships, we could tell which patients had been admitted with amitriptyline overdose: they had the "amitriptyline sign," black discoloration around the mouth due to the ingestion of activated charcoal, which is given to treat the overdose because it binds to the drug, preventing absorption into the bloodstream.

The action of antidepressants on chronic pain is via the pathways that project from the brain down into the spinal cord and can reduce the amount of electrical information signaling harm that is reaching the brain—these are the descending inhibitory pathways we came across in chapter 1. In order to function, these pathways require two chemicals, or neurotransmitters, called serotonin and noradrenaline. When you're depressed these neurotransmitters are at low levels; therefore the pharmacological treatment of depression with drugs consists of trying to elevate those levels. Patients who are in pain are often found to be depressed, and patients who are depressed are more likely to report persistent pain, presumably based on the common biochemical disturbance in these two conditions. In order to activate the descending inhibitory pathways and reduce the patient's perception of pain, the drug needs to be able to raise both serotonin and noradrenaline; that is why drugs like Prozac, which increase only serotonin but not noradrenaline, don't work for persistent pain. Not every patient responds to treatment with antidepressants, and I suspect that the neurochemical disturbance in persistent pain is much more complex than we currently appreciate.

The initial studies for these drugs were carried out on patients who suffer with nerve pain from diabetes or from shingles, because these conditions provide large groups of relatively similar chronic pain patients. Unfortunately, these neuropathic-pain drugs (antidepressants and anti-epileptic medications) are effective in only about one in seven people and they cause harm in about one in fourteen. In addition, they provide only about 30 to 50 percent pain relief, so I wonder how many patients are responding to the treatment and how many are merely recording a placebo response to being given something, when in fact they are activating their own descending inhibitory pathways merely through the act of being looked after.

The newer antidepressant drugs that we use in pain relief include duloxetine, which started its life as a treatment for incontinence, was then used to treat depression, and is now used to treat neuropathic pain, particularly diabetic peripheral neuropathy, a global epidemic as more patients develop diabetes and suffer from the complications that come with this disease. This particular complication is due to the fact that diabetes results in high blood sugar levels, which damage the small blood vessels providing nerves with oxygen; the nerves are then damaged and their electrical activity becomes disordered, causing burning, numbness, and tingling.

When we tell somebody who has no physical damage that we are going to give them an antidepressant to help manage their pain, we may be reinforcing their fear that we consider the pain to be "all in their head." Patients can either get very angry about or completely ignore our advice to take such medications.

We also use drugs such as gabapentin and pregabalin to treat pain, borrowed from the world of epilepsy. As with antidepressants, there is considerable crossover in the biochemical disturbances that cause epilepsy and chronic pain. In neuropathic pain the disturbance is in the orderly arrangement of channels within a nerve. When you damage a nerve in this way it expresses sodium channels in a haphazard fashion, resulting in the firing of the nerve in a spontaneous and unprovoked manner. The patient feels burning, tingling, stinging, and numbness. In epilepsy the aberrant function of nerves may result in seizures or vacant episodes. Gabapentin was so named because it was thought to work on the GABA receptor, which is an inhibitory receptor, but it is now believed that it influences calcium channels. Calcium is important for the release of chemicals that cause nerve excitation, and blocking calcium channels in the spinal cord inhibits the release of the excitatory neurotransmitter glutamate. The hyperexcitable nerves of the dysfunctional pain alarm system are therefore tamed to a degree.

Gabapentin has a number needed to treat (NNT) of seven, which means that approximately one person out of seven that we treat with this medication for nerve pain will have approximately 50 percent pain relief. The NNT is used to indicate the effectiveness of a healthcare intervention. We also talk about numbers needed to harm (NNH), which is the number of people we must give the medication to for one person to develop a significant side effect. The higher the NNH, the better tolerated the drug will be. The NNT for gabapentin is not good, nor for pregabalin, its more expensive cousin. These drugs have been heavily marketed and sold, and we are now in a situation where they are used ubiquitously to treat any pain that goes on for longer than the doctor feels is appropriate. So instead of being used just for chronic nerve pain they are often prescribed for all forms of chronic pain and in the perioperative setting. Yet these drugs make patients foggy and unable to focus, and often make them feel worse. Falling over, confusion, and sleepiness may result from the use of these nerve-modifying drugs. The taming of the pain alarm can even suffocate the patient.

Like ketamine, gabapentin and pregabalin have found a role as drugs of abuse. I am told these drugs have become very popular in prisons because they facilitate the high from other medications. There is now an awakening that we have flooded the market with these antiepileptic medications but have not been mindful about monitoring their untoward effects and their lack of efficacy. But it is always easier to prescribe somebody a drug—and then walk away. With increasing desperation, we continue to borrow drugs from more and more exotic sources for the management of persistent pain.

Doctors are very good at finding, usually accidentally, pharmacological solutions to complex problems. There is an aphorism that for every complex problem there is a solution that is clear, simple, and wrong. Sometimes it is out of desperation in the postoperative setting, where we are managing patients either with difficult-to-treat acute pain or acute-on-chronic pain, that we prescribe antidepressants and antiepileptic medication. The prescribing of a drug is often to manage the distress of the prescriber, to alleviate the feelings of helplessness induced in us by the patient, rather than because there is evidence that the drug will be effective in treating the pain. What all these drugs *will* affect are the patient's emotions and thinking. Perhaps all we do initially is provide the patient with a chemical straitjacket.

I stop medications like gabapentin and antidepressants when they are inappropriately started in the perioperative setting. What prescribers often fail to tell the patient, or the ward staff, is that these medications take approximately six weeks to reach their maximum efficacy. The guidance for neuropathic pain is to start patients on either an antidepressant or an antiepileptic and then, after six weeks, combine the two. There must be millions and millions of unused tablets of gabapentin out there because patients have not been told that for the drug to be effective they have to gradually increase the medication until they experience side effects or reach the maximum allowed dose. They have not been told that the aim is not complete pain relief. So when they don't get complete pain relief, they tell their doctor the medication hasn't worked.

We don't really understand pain as an experience. Much about its pathways and receptors and its uniqueness in individuals remains a mystery. Because of this, the drugs we use are unrefined and unfocused and as a result often produce unwanted side effects. Yet despite our lack of understanding, we have created a medical-industrial complex that sells the idea that most ailments can be cured with medication, from your irritable bowel to your depression. The same is true of pain. There is no marketing strategy for the use of mindfulness or meditation to facilitate pain relief or in the management of preoperative anxiety, even though these may be far more effective remedies than the next poorly understood drug. We are also too reliant on operative interventions, destroying tissues in order to produce functional changes that will hopefully improve pain, despite the fact that we understand pain to be complex and know those functional changes have nothing to do with the outcome for the patient. There is very good evidence, for example, that at least 56 percent of people who have had a technically successful knee replacement continue to experience moderate to severe pain following the surgery. There is as much evidence for the efficacy of conservative treatments that rely on behavioral change such as weight loss and exercise to strengthen quadriceps muscles. But the reality is that most patients prefer to take their chances on the surgery rather than change their behavior and lifestyle.

Medicating pain throughout the surgical process is important, but it is not the whole picture, and several of the medications we use are flawed and dangerous. We often collude with patients in terms of the success of pharmacological management of chronic pain and acute pain. As doctors, we see patients on an irregular basis in the chronic pain setting, and so prescribing a medication is an easy undertaking. What we don't see is the patient's gradual cognitive decline due to the medication; we don't see the patient falling over and fracturing their hip and coming into the hospital and getting pneumonia and dying. The hip fracture and the pneumonia are managed by orthopedic teams, but the link between the trauma and the high-dose opioid that made them fall over in the first place is not highlighted or revealed to the opioid prescriber. Until now.

CHAPTER FOUR

The Line between Pleasure and Pain: Opioids and Addiction

Among the remedies which it has pleased Almighty God to give to man to relieve his sufferings, none is so universal and so efficacious as opium.
—Thomas Sydenham ("the father of English medicine")

I started to take codeine and paracetamol. At first it was only once or twice a day, but as the back pain got worse and I struggled to do things on a day-to-day basis and I became more miserable and worried, I found myself taking it every three to four hours—it made me feel better. There were times when the pain became worse and I would not be able to get out of bed and when I went to my GP, he gave me tramadol and he also started me on fast-acting liquid morphine. He told me the medication would help but it didn't, and I went back to him confused about whether to take the codeine, the tramadol or the morphine. The morphine seemed to help the most. It made me feel spaced out during the day though. I tried some injections into my spine, but the back pain never stopped. I started to have the morphine every four to six hours. At first, I used a spoon but eventually I started to just drink it straight from the bottle.

As the years went by, I continued to take the morphine, drinking straight from the bottle; I felt better, less weary and depressed when I took it—at least for a while. I also started to shake if I ran out of morphine. I started to sleep in the middle of the day to help with the pain. The pain clinic told me that I was taking too much morphine and said that they wanted to put me on a patch. I tried the patch, but it didn't help. I continued to take the morphine out of the bottle. I felt my world getting smaller and smaller—just me and my morphine. My GP then started me on long-acting morphine tablets. I didn't find this very helpful and continued to take the liquid morphine.

The pain clinic tried to help by giving me exercises to do and teaching me how to cope with the pain. I still had the pain though and so I asked my GP to increase the long-acting morphine because I was taking more and more of the liquid morphine. Six years after I started taking codeine I was now taking 20 mg of long-acting morphine twice a day and using the liquid morphine every four hours. Nobody could tell me what was wrong with my back and why I had pain. My GP then increased my long-acting morphine to 30 mg. He also tried me on another medication, an antidepressant and anti-epileptic as well as patches of local anesthetic, but nothing helped.

Eventually I was taking 100 mg of the long-acting morphine and 80 mg of the liquid morphine every day. The GP tried another patch, which he said was equal to about 400 to 600 mg of morphine a day. The patch couldn't stay stuck to my skin, so he switched me back to the liquid morphine. I noticed that my teeth started to rot. I think it was because of the sugar in the liquid morphine. I was using a liter bottle of morphine which would last me about a week or two—sometimes less.

The morphine never helped with my pain. I still had the pain, but I found that I didn't really care about the fact that I had pain. I would be able to sit quietly in the corner and just feel less bad about having pain. The long-acting morphine didn't really help. I suppose it was there in the background doing something, but I never really found that it made me feel as if I could do more things around the house. The liquid morphine was much better and gave me the space to feel a bit less unhappy. It would only last a short period of time and whenever I was really unhappy or distressed, I would take a swig from the bottle. This made me feel a little bit better. The morphine made me constipated, but I used to take laxatives. Sometimes I would have to take the stool out of my bottom with my finger. I never felt particularly well when I was on the morphine, but I felt worse when I wasn't taking it.

This testimony came from a patient, Diane, who shared with me this story about her relationship with morphine. (We will hear more from Diane in chapter 6.) She is not alone in her struggle with morphine dependency.

———————

Human beings have a long and complex history with opium, dating back to around 5000 BCE. While much of its history is medical, opium has also been used recreationally for centuries. In fact, Samuel Taylor Coleridge is thought to have composed the poem "Kubla Khan" as a reflection on his experience with opium. Opium is derived from the opium poppy (*Papaver somniferum*) and is contained within the dried latex obtained from this plant, 12 percent of which is made up of morphine, which can be processed to produce heroin. The latex also contains codeine and thebaine, which form the basis of the production of semi-synthetic opioids such as oxycodone and hydromorphone.

Opium was widely cultivated in the ancient world. The Sumerians grew opium poppies as early as 3400 BCE and called it the "joy plant." The Egyptians cultivated opium around 1300 BCE in the ancient city of Thebes, which is why the active compound in opium is named thebaine. The ancient Egyptians restricted the use of opium to priests, who were the gatekeepers of healing and death. Their use of the substance was medicinal; the Ebers Papyrus (c. 1550 BCE), one of the oldest medical papyri discovered, includes a description of opium as a remedy to prevent excessive crying in children, for treating breast infections, and for reducing pain during surgical procedures.

The ancient Greek and Roman civilizations also cultivated the plant, and culling knives used to harvest opium have been found in Cyprus thought to date back to 1100 BCE. The trade in opium was widespread in the areas surrounding the Mediterranean Sea, where the oldest seeds from the poppy plant have been found. Hippocrates denounced the magical properties associated with opium but recognized its usefulness for treating pain and disease, for inducing sleep, and even for euthanasia. There is evidence that the Greeks used opium in a variety of forms, including inhaling it as a vapor and using it as a suppository. Homer writes in the *Odyssey* that Polydamna, wife of the Egyptian king Thonis, gave Helena, the daughter of Zeus, the mysterious drug nepenthe, which could be slipped into wine and had the power to banish all painful memories. Menelaus and his companions are given nepenthe to forget the horrors of the Trojan War when these painful memories are recalled by the arrival at court of Telemachus, the son of Odysseus, who fought alongside Menelaus. Homer says that nepenthe was grown in the Egyptian fields, and the word "nepenthe" can mean "no grief" or "no sorrow" as well as "no pain," "painless," or "canceling all pain." It seems likely that the drug being referred to here is some form of opium. Opium was also used as a constituent of the hemlock cup used by the Greeks to ensure a painless death, and the description by Plato of the death of Socrates is consistent with the notion that the fatal drink contained opium.

Although the use of intoxicating substances was banned in Islam, there is evidence that the use of opium was common in the places conquered by Muslims between 600 and 1500 CE, particularly Turkey, where it was drunk as part of a black tea. Arab traders introduced opium to China during the Tang dynasty (618–907 CE) and to India in roughly 700 CE, and the drug was used by Persian physicians to facilitate anesthesia for surgical procedures. Persian physicians also recommended opium for the management of melancholy, and it was used as an everyday remedy for a variety of aches and ailments in the absence of access to a physician. The effect of opium at the time of the Persians was already known to include pain relief, suppression of coughing, effects on the mind, and inhibition of the ability to breathe, as well as disturbance of sexual function.

Opium arrived in Western Europe with soldiers returning from the Crusades between the eleventh and thirteenth centuries, but was a taboo subject because the zealots of the Inquisition believed anything from the East was from the devil. And so the drug was lost to European medicine for more than two hundred years. The Swiss physician, astrologer, and chemist Paracelsus (Philippus Aureolus Theophrastus Bombastus von Hohenheim, to give him his full birth name), who established the role of chemistry in medicine, encountered opium on his travels in the Middle East in the early sixteenth century. Paracelsus held the view that a doctor must be well-traveled and that knowledge is experience, and he is quoted as saying, "A doctor must seek out old wives, gypsies, sorcerers, wandering tribes, old robbers, and such outlaws and take lessons from them." He eventually returned to Europe and introduced opium into Western medicine. He is credited with the invention of the drug laudanum, whose name means "worthy of praise," which consisted of opium mixed with amber, crushed pearls, musk, and other substances, including saffron.

Laudanum remained relatively unknown until an English physician, Thomas Sydenham, proposed his own proprietary formulation in the seventeenth century and recommended its use for a variety of ailments. In a world of appalling living conditions, where individuals were beset with diarrhea due to cholera, with persistent coughs and various other pains, an opiate-based medicine represented a panacea, and Sydenham's laudanum retained its popularity well into the eighteenth century. Opium is an effective treatment for cholera and diarrhea because it causes constipation; it was also used for joint pain and difficulty with sleeping. The only medical alternatives to laudanum at this time were poisonous substances such as arsenic, mercury, and drugs designed to induce vomiting.

Where there was psychological disequilibrium, opium found a place. The effects of opioids in the form of alleviating anxiety and increasing dopamine levels with consequent production of euphoria resulted in its popularity among physicians who even recommended it to people without any maladies in order to equilibrate the internal environment of the body. Throughout the eighteenth century, the sedative effects of opiates were used to treat patients with psychosis, a more humane treatment than the restraint methods that until then had been the predominant form of psychiatric treatment, based on the idea that people with mental health problems were insensitive animals. In the US, opioid prescription increased in the nineteenth century, when it was ubiquitously given to women suffering from complaints attributed to their gender, including menstrual pain and hysteria.

―――――――――

Opium also played a major political and economic role in the history of China. One of the earliest records of opium use in China dates back to 1483, in which it is described as aiding masculinity and enhancing the art of sex through delayed ejaculation. The introduction of tobacco smoking to China by the Portuguese in the 1500s is perhaps responsible for the use of opium becoming more extensive, because opium was then mixed with tobacco, transforming the drug's use from an indulgence of the elite to a widespread phenomenon. The ruling elite feared that groups of opium smokers congregating in the southern provinces of China were plotting against the emperor, but given the calming effect of opiates, this fear of a revolutionary predisposition is likely to have been unfounded. Nevertheless, the social ills caused by opium use as well as the authorities' fear of these groups of "opium-eaters" led to decrees banning the drug. Its prohibition began in 1729 in China, and it was completely banned from 1799 until 1860.

By the eighteenth century China was known as the silver graveyard of the world because the West, particularly England, paid for Chinese tea, silk, and porcelain—but mainly tea—in silver. The English traded opium illegally to the Chinese in order to support the British addiction to tea. In 1839 the First Opium War was triggered when the viceroy of

Huguang, Lin Zexu, destroyed twenty thousand chests of opium supplied by the British as payment for tea—burning the drug with fire, salt, and lime and flushing it into the sea. The event took place on June 3, the date still commemorated as Antidrug Day in China. The Treaty of Nanking was signed when the Chinese were defeated, and as part of the agreement the British were given the barren rock in the ocean that would become Hong Kong.

The imperial commissioner of Canton was determined to stamp out the opioid trade, however, and closed Canton to foreign traders. The ensuing conflict with the British and the French who sought to regain this gateway to China resulted in the Second Opium War, which lasted from 1856 to 1860. This time the Chinese defeat brought the forced legalization of the opioid trade, the expansion of trade in cheap labor, and exemption of foreign imports from internal trade tariffs. China began massive domestic production of opium; it is estimated that by 1905 approximately 25 percent of the male Chinese population were regular consumers of the drug, and by 1906 13.5 million people across the country were consuming thirty-nine thousand tons of opium per year. The Second Opium War opened trade to the West; five more Chinese ports were opened to direct foreign trade rather than having to engage with the government. It can therefore be argued that opium is the reason China did not continue to be an isolationist empire retreating behind the Great Wall. It is also easy to understand why today the Chinese might feel aggrieved at being lectured by Western powers about human rights violations and the production and export of Chinese-made fentanyl, a potent synthetic opioid used by anesthesiologists in hospitals for pain relief after surgery and in a slow-release patch formulation for chronic pain, and which is widely regarded to be fueling the current opioid crisis in the US.

In the latter part of the nineteenth century, there was widespread movement of Chinese and Indian workers to other parts of the world, particularly the western United States to engage in mining, railroad construction, and agriculture. These workers brought with them the habit of opium smoking and established "opium dens" in Chinatowns. This was particularly the case with Chinese immigrants moving to the United States, and restrictions on who could use opium were soon introduced, for fear that the local populations would become addicted. The Chinese migration in 1849 to San Francisco in California led to the formation of the International Opium Commission, a body designed to instigate the global prohibition of drugs.

Opioids are naturally occurring and synthetic substances that bind to opioid receptors within the body. Opiates are obtained from the opium poppy, *Papaver somniferum*; morphine, codeine, and thebaine are active compounds derived from the plant. Morphine (named after Morpheus, the ancient Greek god of dreams) was first synthesized in Germany in 1804 by Friedrich Sertürner and was produced in industrial quantities in Germany in the 1820s. The development of the syringe by Alexander Wood in 1855 resulted in the ability to administer morphine easily and frequently, perhaps contributing inadvertently to the overuse of opiates. The use and administration of opiates were not regulated in the US prior to the Harrison Narcotics Tax Act of 1914, and individuals could obtain morphine for a variety of ailments. The Sears and Roebuck catalogue offered morphine and a syringe for $1.50. Alcohol-based treatments were also widespread—if it made you feel good, it must be good for you, seemed to be the prevailing view. "Headache powders" containing opium were available over the counter and were self-administered without recourse to medical assessment or diagnosis. It wasn't until the mid-nineteenth century that doctors began to realize that the regular use of morphine had negative effects.

Heroin, which is a derivative of morphine, was first chemically synthesized by Charles Romley Alder Wright in 1874. At the time, morphine was a popular recreational drug and was widely available, finding its way into many patented remedies, such as Mrs. Winslow's Soothing Syrup (containing morphine and alcohol), recommended for fussy children, and formulations for menstrual cramps. The Bayer Company of Germany wished to find a similar but less addictive alternative and in 1895 introduced their new drug. Under the trademark "Heroin," based on the German word for "strong" and "heroic," it was sold as an over-the-counter remedy for cough suppression. Human beings don't learn at all: Purdue Pharma would make the same mistake marketing oxycodone (a synthetic opioid) as a less addictive alternative to morphine some hundred years after Bayer first marketed Heroin.

By 1910 there was an epidemic of heroin use in the US. Working-class Americans had learned to crush the pills into powder and inhale it to achieve a concentrated euphoria. In the mid-twentieth century heroin had become the recreational drug of choice among jazz and rock musicians. The anthropologist Michael Agar talks interestingly about heroin being the ideal drug for New York, a city that delivers a constant information overload. He refers to heroin as the audiovisual technology that helps manage that overload by dampening it across the board, allowing users to focus on one small aspect of it, on a scale that the human perceptual equipment was designed to handle. Perhaps this is the reason chronic-pain patients sometimes find morphine useful: it dulls and distracts.

Heroin was eventually banned by the League of Nations in 1925; this led to the production of related drugs such as oxymorphone and hydrocodone in massive quantities to close the gap in the market. In the US the Committee on Drug Addiction, formed under the banner of the National Research Council in 1929, supervised research into the morphine molecule in order to find a nonaddictive, safe alternative. As a result of this research several different opiate analgesics were derived, including oxycodone and methadone, but the strong, nonaddictive alternative to morphine that researchers set out to develop has still not been created.

In the UK, the Department of Health's Rolleston Committee report established guidelines in 1926 for the use and prescription of diamorphine (the chemical compound of which Heroin is a trademarked product), and heroin continues to be widely used in palliative care in the UK. Some users believe that it causes more euphoria than other kinds of morphine, which may be related to the way it is metabolized to morphine in the body. It is more potent than morphine because it is more fat-soluble and therefore more rapidly absorbed by the brain. Heroin also continues to be a major drug of abuse in the UK.

In 1899, shortly after marketing Heroin, Bayer synthesized aspirin; it became an over-the-counter drug worldwide after 1917, and for the majority of the global population became the drug of choice to treat pain, replacing opiates. For cancer pain, however, opiates remained essential. There was growing criticism of doctors who withheld morphine from patients suffering with cancer, but these doctors were conflicted between relieving the individual's pain and risking possible addiction. Opiates were therefore withheld until the last few weeks of a patient's life.

Dame Cicely Saunders (1918–2005), regarded as the founder of palliative medicine in the UK and responsible for establishing the first purpose-built hospice in 1967, was a strong proponent of using opiates for cancer pain. She developed the Brompton Cocktail, a mixture of morphine, cocaine, and gin, in order to ease the suffering of her patients and improve their death. She coined the term "total pain," which she described as a clinical phenomenon that compounded physical and mental distress with social, spiritual, and emotional concerns. Studies done at St. Christopher's Hospice by Robert Twycross demonstrated that morphine was superior to and more efficient than heroin and did not bring about tolerance or addiction in cancer patients even with long-term use.

In 1982 the World Health Organization's Cancer Unit convened to produce the WHO ladder, a set of guidelines for physicians managing cancer pain. It recommended that they prescribe analgesics on a regular three-step schedule and measure and adjust the dosage according to the patient's pain at each of the three steps—starting with anti-inflammatory medications like aspirin, moving on to codeine, and finally morphine. Some physicians recommended the ladder to patients with chronic noncancer pain, but there remained concerns about prescribing opiate medications in this situation. When we ask medical students today how to manage any kind of pain their answer is "the WHO ladder," so powerfully has this approach been marketed, even though it is rarely appropriate for acute pain and is inappropriate for chronic noncancer pain.

In 1986 the World Health Organization published its "Cancer Pain" monograph, and the management of cancer pain became a priority in many countries; however, some countries today still suffer from poor management of cancer pain with opioids, due to the pejorative connotations of opioid use. In the 1990s several publications looking at the undertreatment of pain asked why opioids were solely reserved for cancer pain. The work of Kathleen Foley in the 1970s at Memorial Sloan Kettering Cancer Center in New York was geared toward understanding the use of opioids in cancer pain. She believed the censure against opioids made little sense. She was influenced by the work of her mentor, Raymond Houde, who conducted crossover studies of analgesics in cancer patients, showing that morphine was superior to other forms of pain relief. Houde's crossover studies indicated that it was a misconception that morphine-type drugs enslaved, demoralized, and led the unwitting patient down the path to addiction. Foley was also influenced by the work of Saunders and Twycross, who were using regular heroin dosages to free patients from what they perceived to be the twenty-four-hour nightmare of constant pain. They viewed the patient's pain as being of greater harm than any potential risk of developing dependency on morphine.

The push from cancer pain specialists to treat their patients with opioids spilled over into the treatment of chronic noncancer pain, leading to an indiscriminate and unreflective assessment of the differences between cancer and noncancer pain. Foley published articles in 1981 and 1986 indicating the low incidence of addictive behavior in small groups of cancer and noncancer patients. The first piece was published as a one-paragraph letter without detailing any scientific method as to how the conclusions were reached. The authors asserted that there was a 0.03 percent risk of addiction for patients receiving opioids for acute pain. The other study looked at thirty-eight patients and was a retrospective case review. According to this evidence only two of the thirty-eight patients with chronic pain developed misuse or abuse when receiving opioids. This inadequate evidence became the scientific basis for a twenty-year campaign for the long-term use of opioids in cases of chronic noncancer pain, led by Foley and her colleague Russell Portenoy. Dr. Portenoy said in a 2010 videotaped interview, "I gave innumerable lectures in the late 1980s and 90s about addiction that weren't true" and that it was "quite scary" to think how the growth in opioid prescribing driven by people like him had contributed to soaring rates of addiction and deaths due to overdose. It is estimated that in the past twenty years approximately $8 billion was paid in inducements to physicians to expound on the safety and benefits of opioids in the US. Portenoy was sponsored by Purdue Pharma and other companies to promote the use of opioids in chronic noncancer pain but has denied that these financial relationships influenced his views.

Chronic nonmalignant pain and cancer pain are complex biopsychosocial phenomena. The difference between the two is that cancer pain in the palliative care setting has a terminal endpoint for the prescribing of opioids because the patient usually dies. Interestingly it is now known that high-dose opioids used to treat cancer pain may in fact affect the absolute longevity and quality of life of the individual. This may be because of the immunosuppressive effect of opioids, which accelerates the cancer.

Unexplained pain is particularly hard to treat. In the 1920s these patients were regarded as deluded or malingering or were accused of being drug abusers. Unfortunately, despite what we now know about pain, this is still sometimes the explanation offered to patients who suffer with chronic noncancer pain. In the 1970s there was increasing criticism of the failure of clinicians to treat patients with adequate doses of opioids, and in the 1980s a widespread belief developed that the use of opioids for therapeutic indications rarely resulted in addiction. The evidence base for this assertion was the two small retrospective publications by Foley from 1981 and 1986 on studies conducted with cancer patients. The use of opioids for chronic noncancer pain was therefore based on two small studies without any prospective follow-up regarding safety issues or extensive studies randomizing groups of patients with pain conditions and testing whether opiates were of benefit.

In 1995 the American Pain Society launched a campaign labeling pain "the fifth vital sign." The aim of this movement was to encourage the standardized evaluation and treatment of pain symptoms. The US Veterans Health Administration lent support to this campaign, and in 1999 they too adopted the idea of pain as the fifth vital sign. Further government input emphasized the need for assessment of pain. The change in how opioids were seen in the management of pain was aided by the Drug Enforcement Administration, which promised less regulatory scrutiny over doctors who prescribed opioids; these doctors were also promised support by the Federation of State Medical Boards. Hospitals implemented these changes and instituted systems for pain assessment. Pain management was mandated, and if benchmarks were not met, a hospital was likely to be penalized. American hospitals therefore invested heavily in opioid therapy in order to receive better satisfaction ratings from their patients. The same revolution in the aggressive management of pain was seen in the UK with the start of inpatient pain services to manage acute pain after the report "Pain after Surgery," published in 1990, highlighted the inadequacies of postoperative pain management in UK hospitals.

Pharmaceutical companies got on board and heavily pushed the use of opioids, aggressively marketing them under the banner of being a humane treatment option. Trainees in pain medicine and other specialties were taught to rely on opioids for pain treatment. Newer formulations such as oxycodone were introduced and were marketed as having a lower likelihood of abuse. In fact, however, oxycodone—nicknamed "hillbilly heroin" in some states in the US—has been heavily abused. Between 1997 and 2002 oxycodone prescriptions increased from 670,000 to 6.2 million in the US, and overall opioid consumption rose from 46,946 kilograms in the year 2000 to a peak of 165,525 in 2012.

Initially the increase in opioid consumption did not appear to increase the risks of complications in the perioperative setting. However, concerns soon arose after an almost twofold increase in incidents of oversedation due to opioids; then a link was established between overaggressive pain management and this substantial increase in oversedation associated with fatal respiratory depression. Between 1999 and 2016 opioid sales quadrupled and mortality rates in both men and women as a result of opiate use also quadrupled. Evidence of increasing pain, worsening disability, hormonal problems, and increased anxiety and depression have also been found to be related to opioid use. Yet despite this increasing mortality rate and the ubiquitous use of morphine to treat chronic noncancer pain there has been no Level I evidence (a randomized controlled trial giving one group the opioid and the other a placebo) evaluating the efficacy or safety of the use of opioids in reducing pain and improving function in chronic noncancer pain.

During the 1990s the aggressive marketing in America of OxyContin (the brand name of oxycodone) by Purdue Pharma intersected with the trafficking of heroin in cities across the Midwest. OxyContin was said to have a low risk of addiction because it was a prolonged-release formulation, but the situation was a repeat of the crisis a hundred years earlier, when Bayer produced Heroin as a less addictive alternative to morphine. Users soon realized that they could crush or snort the OxyContin to achieve a high. Liberal prescriptions were written by overworked GPs with no access to multidisciplinary pain clinics. Once recognized, the misbranding of OxyContin by Purdue Pharma resulted in their having to pay fines totaling $634.5 million in 2007. They were accused of intentionally downplaying the risk of addiction and misleading physicians and the healthcare industry by overstating the benefits of opioids for chronic pain. Purdue therefore agreed to cut its sales force in half and stop promoting opioids.

In the late 2000s the price of OxyContin and other prescription opioids increased, becoming too expensive for many users, particularly in Middle America. A good proportion of these people therefore switched to the less expensive black tar heroin, exacerbating the country's longstanding problem with heroin abuse. Over time middle-class American adolescents and young adults graduated from OxyContin to heroin. As a consequence, in the past decade, the number of deaths due to drug overdose has increased 137 percent across America and overdoses specifically involving prescription opioids and heroin have increased by 200 percent. In 2016 sixty-four thousand people died from drug overdose; of those, forty-two thousand were opioid deaths, representing a 20 percent increase from 2015.

We are now in the grip of an opioid epidemic—again.

Due to the staggering increase in opioid-related deaths in the US, the epidemic was declared on October 16, 2017, and is considered a public health emergency. Illegally manufactured fentanyl represents the greatest contribution to this increase.

The opioid epidemic has even reached popular culture. Fictional star of page and screen Jack Reacher has been exposed to it, and David Baldacci has written a novel that depicts opioid misuse in Middle America. The BBC has done programs on opioids and their effect on society.

The opioid epidemic demonstrates our destructive tendency to treat a complex experience such as pain with medication in the absence of a clear understanding of the myriad chemical interactions that underpin this phenomenon. A hammer has been used to crack a nut and, even though sometimes the nut is revealed, it is no longer edible.

But what is the alternative to a biomedical approach spearheaded by doctors armed with medicines, scalpels, and needles? In America alone 25 million adults suffer with chronic pain and 23 million with severe recurrent pain. In the UK, one in seven adults, possibly as many as 10 million, are thought to suffer with chronic noncancer pain. This leads to disability and loss of work productivity as well as compromised quality of health and life. Clearly a solution is needed. Rather than relying on opioids, we need to adopt a biopsychosocial approach to the management of pain, including pain management programs employing cognitive behavior therapy, relaxation and coping skills, and even self-hypnosis. Unfortunately, insurers regard this approach as being too costly; prescribing an opioid appears, on the surface, to be a much simpler solution.

The International Association of the Study of Pain declared in 2010 that the relief of pain and suffering is a moral duty of physicians and that access to adequate pain relief is a fundamental human right— but we have now swung the other way, seeing the simplistic assessment of pain severity in terms of a score from 0 to 10 as dangerous because it leads to excessive prescribing of opioids when you are treating a number, with no emphasis on functional goals. What should be assessed is not the pain score but the impact of pain on function. I would, for example, say to a patient who has fractured ribs that the aim of pain relief is to facilitate deep breathing and coughing to prevent a chest infection, rather than treating a number that has been ascribed to the level of pain.

New standards in pain management now include recommendations relating to psychosocial risk factors that may affect the self-reporting of pain, the setting of realistic goals, and developing treatment plans that focus on the ability of a medication to increase physical functioning. The guidance for assessing whether a pain medication should be used and the outcome measures that are to be assessed now follow the recommendations of the Initiative on Methods, Measurement, and Pain Assessment in Clinical Trials (IMMPACT). The effect of the medication should be evaluated in terms of (1) pain, (2) physical functioning, (3) emotional functioning, (4) participant ratings of improvement and satisfaction with treatment, (5) symptoms and adverse events, and (6) participant disposition (for example, adherence to the treatment regimen and reasons for premature withdrawal from the trial).

———————————

Chronic pain is often found to develop in patients who have undergone repeated abdominal operations and episodes of inflammation, resulting in a hyperactive and sensitive pain alarm system. These patients develop what is called visceral neuropathic pain or chronic abdominal pain. The difficulty with pain that comes from an organ as opposed to pain that comes from a muscle or a ligament is that the nerves to an organ are very poorly localized and hitchhike with nerves that go to muscles. If you have ever experienced diarrhea you will remember the unpleasant colicky sensation from bowels that are behaving abnormally and how the pain seems to be everywhere. That constant pain is accompanied by an inability to eat. Patients who have chronic abdominal pain therefore have a miserable existence, and their distress usually leads to the prescribing of high-dose opioids.

By the time pain specialists such as myself are called in to see patients with intestinal failure, the patient is usually already on high-dose opiates and we are asked to reduce the patient's use of these drugs, which is like taking a teenager's phone away from them—forever. Often these patients have developed habits associated with their opioid use, which might include wanting to be given only injections even after they have started to eat and drink again. The explosion in the types of opiates and the variety in routes of entry means that patients who cannot eat and drink are often given these newer formulations, such as lollipops, sprays, and patches, which can reinforce addiction. Lollipops containing high doses of fentanyl which are absorbed through the mouth and intranasal opiates that are inhaled are two routes that favor the development of addiction because the nature of the absorption produces rapid and profound effects.

When we see these patients we introduce the concept of chronic pain and begin a conversation about the harm that opiates can cause. But somebody who suffers with chronic abdominal pain finds it very hard to accept that what we are offering is better in the long term than the temporary but immediate escape offered by an opioid. Battles therefore rage around opioid reduction. I sometimes wonder when I go on these wards whether it is not kinder to just allow patients to have their opiates.

Our understanding of what goes wrong in the way nerves behave in pain without injury is undeveloped; drug companies therefore do not have a precise target to aim for when trying to develop medicines for pain. Pain, however, is an emotional experience, and seeing somebody in pain evokes in healthcare workers as well as loved ones the desire to rescue. This need to alleviate pain in others is what led to the opioid crisis in the first place, when we decided to champion the idea of managing this experience and, through our hubris, felt that we could do so medically, ignoring the fact that chronic pain is an experience and needs to be approached from a biopsychosocial point of view.

We have still not learned our lesson and are once again approaching the same experience and trying to modify it with another pharmacological agent in the form of cannabis. Industry has got on board and cannabis stocks have soared. So instead of calmly assessing where this medication fits into the management of specific pain conditions and then carefully conducting randomized controlled trials in order to assess which patients benefit from what dose and what formulation of the drug, we are steaming ahead again. When you take one tool away from doctors, in order to sustain their position as providers of healthcare they move on to another medication that can be controlled and prescribed. There have so far been only twelve randomized controlled trials on cannabis in the past five years, and most of these studies have indicated that cannabinoids are not effective in the management of neuropathic pain. Yet there is a real danger that we will supplant one mind-altering substance with another, all because we cannot appreciate that health and well-being are about more than pharmacological agents that can simply be packaged and sold.

Over the years we have used, abused, and promoted the use of opium. But while opium has been used for pain, that is not the predominant narrative in the relationship we have had with the drug. We have used its tranquilizing, constipation-inducing, cognitive and behavioral modification effects far more. So it has become clear that using opium for pain is about managing the behavior of the sufferer rather than arresting the pain alarm or treating the cause of the pain.

Our dysfunctional history with opiates perhaps reflects our struggles with the slings and arrows of life itself. The deaths caused by opiate use have been termed "deaths of despair," implying that opioid abuse reflects deeper societal issues alongside our complex relationship with understanding pain. Pain may be due to physical injury, but the discomfort we feel when we are emotionally distressed is no less unpleasant than a broken bone. When life becomes too much and we feel overwhelmed, we may try to cope by pushing these feelings down with alcohol, sugar, opioids, and cannabis. These drugs narrow the way we see the world, causing our brain's ability to perceive the world to shrink, lessening our anxiety and despair.

The relationship we have had with opiates is a cautionary tale. The fact that physicians working in cancer pain management, who heroically want to save the patient from suffering and are involved in the care of the dying, were able to state with very little supporting evidence that morphine is perfectly safe in managing pain in a non-terminal condition is a worrying example of how a therapy can be introduced despite little evidence of its benefits and centuries of recorded harm. It is extraordinary that the lessons of history have not been heeded with regard to our relationship with opiates. There is so much repetition to be seen in the introduction of the Harrison Narcotics Tax Act in 1914, the Opium Wars, and the introduction of Heroin a hundred years ago, reflected more recently in the wars in Afghanistan, the introduction of oxycodone, and our current opioid epidemic.

In the *New York Times* in 1911, the US opium commissioner Dr. Hamilton Wright was quoted as saying:

> *Of all the nations of the world, the United States consumes the most habit-forming drugs per capita. Opium, the most pernicious drug known to humanity, is surrounded, in this country, with far fewer safeguards than any other nation in Europe fences it with. China now guards it with much greater care than we do; Japan preserves her people from it far more intelligently than we do ours, who can buy it, in almost any form, in every tenth one of our drug stores.*

Those who ignore the lessons of history are doomed to repeat its mistakes.

━━━━━

When it comes to treating pain, we sway from one end of the spectrum to another, just like skirts get longer and shorter with no real rhyme or reason. All the while, those who prescribe opioids probably don't take them, and those who take the opioids are probably never given enough information about the harm the drug is causing. We would rather provide a single drug—and now cannabis is rearing its head—than admit that people need to understand and learn to manage their pain. And we would rather prescribe a medication than invest in psychological therapies or physiotherapy services.

One of the main functions of the pain clinic I work in is to help people reduce their opioid use. The conversation with the patient who was prescribed opioids is difficult. How do you say: Well, the drug that has been given to you, that you are now on, and which has no evidence base, is poisoning you. The contract between patient and doctor when starting opioid treatment should include not merely writing the prescription but also documentation of a discussion on the risks and benefits of the therapy. Historically, however, this has not been the case. GPs who have for years prescribed opioids like Smarties are now asking us to position ourselves as an opioid-reduction clinic. And because pain clinics are under threat since pain injection therapies are being stopped, we have decided that this is probably the way forward.

There may be a role for the use of opioids in chronic pain, but it is not a role that we have studied, and therefore we do not have the necessary information to recommend its use. We have evidence regarding the effective use of opioids in the management of acute postoperative pain but increasingly realize that we should be more cautious in their use, monitoring for signs of addiction and limiting the amount prescribed.

In medicine these days we act as if we've cracked it. We believe that because we are so technologically advanced and have these magnificent temples called hospitals, we have all the answers. The reality is far different; we swing from one failed pharmacological or surgical therapy to another without appreciating or admitting that the solutions to healthcare problems in the first world are complex. Poor lifestyles cannot be medicated, and wellness cannot be prescribed.

Until patients challenge doctors and engage much more in the decision-making around their situation, we will continue in our current vein, and in the next hundred years we will probably swing back to prescribing opioids ubiquitously—or perhaps the new morphine will be cannabis.

CHAPTER FIVE

Pain with No Injury

U sually pain goes away, not dramatically, like an unwanted guest you kick out of your house, but quietly. At the end of its time with you it becomes an itch at the edge of your perception, and then eventually it is no longer present at all. The alarm switches off because your body has healed. There are times, however, when, despite all the tissues knitting together, pain persists.

Most people think of pain as a symptom of injury or harm, and the main preoccupation of the doctor is to find out what the cause of the pain is. Doctors often employ an approach to medicine that involves gathering many pieces of information and whittling them down to a single diagnostic label, aiming to reduce the complexity presented by individuals with persistent pain and a multitude of symptoms down to a single point of origin—a pain generator.

But pain is a very complex biological alarm system. It is biologically ancient, with pathways that are hardwired into multiple areas of the brain, and it is not as specific a guide to danger as we like to think it is. Dysfunction of the organs, for example the heart or the pancreas, results in poorly localized pain to the left arm and the middle of the stomach, respectively. The specificity of pain as a symptom is therefore often overestimated. A straightforward diagnosis—a pain generator—cannot always be found. The tools doctors employ to reach such a diagnosis have traditionally been taking a history of pain, ordering tests, and performing a clinical examination. In the past medical students were encouraged to use the mnemonic derived from the name Socrates when assessing pain: site, onset, character, radiation, associations, time course, exacerbating/relieving factors, and severity.

Medical students today, in contrast, are increasingly taught the Calgary Cambridge method of consultation. First the doctor will make every attempt to build rapport by asking open questions and employing active listening techniques, placing the patient at the center of care. They collect information in order to assess both the disturbance in the person's physiology (the disease) as well as an understanding of the psychosocial factors that may impact the disease. This change in how doctors assess patients is rooted in the understanding that *disease* and *illness* are two very different concepts: disease is what the doctor says the patient has, and illness is what the patient suffers with. Doctors assess and treat for disease, while patients report illness. The disease is determined through a process similar to the old method, by taking a history, performing an examination, and ordering tests to narrow down the possibilities with regard to what is different from normal or not functioning properly. The concept of *illness* attempts to understand the disease within the context of the patient, because this has implications for how successfully that condition may be treated. Unfortunately, most doctors still operate a reductionist model and ignore the psychosocial aspect of a patient's health.

Lower-back pain, for example, is a universal human affliction and is often attributed to dysfunction of various components of the spine, which is a very complex structure consisting of vertebral bodies stacked one on top of the other. Various tests have been devised by doctors in order to mimic the pain the patient presents with and therefore identify the pain generator—the source of the problem and the bit of the patient's back that can be fixed with screws and rods, cut out with a scalpel, or made numb by burning its nerve supply. This follows the reductionist trap of trying to refine a disease and define a disease process. But the spine is not a structure that is easily defined in such terms, and the innervation of the spine is complex; one nerve may in fact give sensation to multiple structures.

Nociception is the detection of harmful stimuli by specialized receptors and nerve endings, and this information that damage has occurred to the body is transmitted to the spinal cord and brain. Pain is the way the individual *perceives* this harmful information. Pain occurs when the person filters the information about harm through their past life experiences and current psychological makeup. The information about harm is also filtered through the cultural and social lens of that person, and the presentation to the GP with symptoms of pain will very much depend on the individual's psychosocial context.

Thinking about humans as having a body separate from their mind rather than being an integrated entity leads to a poor assessment and insufficient management in healthcare. When the body is seen as a machine and the focus is on looking for a disease, a disruption of bones and ligaments, or dysfunction of organs, the impact of psychological and social factors on both the initial reporting of symptoms and the subsequent response to treatment is ignored, leading to inadequate management and a lack of progress in improving function.

As we have explored, pain is an abnormal sensation emanating from the body and is interpreted through the person's own unique experiences and understanding of normal. Pain is the final expression of what the individual understands about their body and their tolerance for discomfort. Pain is culturally determined: where pain is due to an obvious cause, such as a broken limb or a musculoskeletal strain, it is relatively easy to understand; where pain occurs insidiously and progressively it is more difficult to understand. We are as a species meaning-makers, and if our doctor can't explain our pain we will develop our own narrative rooted within our cultural context. So what happens when there is pain and illness that has persisted beyond the normal period of healing, but no evidence remains of the initial disease, injury, or ongoing tissue damage? At this point, a patient is said to suffer from chronic pain.

Chronic pain is divided into two categories. Primary chronic pain syndromes include fibromyalgia syndrome, complex regional pain syndrome, and nonspecific lower-back pain. Primary chronic pain is defined as pain in one or more anatomical regions that persists or recurs for longer than three months and is associated with significant emotional distress or functional disability (that is to say, it interferes with the activities of daily life and restricts participation in social roles) and that cannot be better accounted for by another (nonprimary) chronic pain condition. This is a new definition, which applies to chronic pain syndromes that are best conceived as health conditions in their own right. Secondary chronic pain is persistent pain whose initial cause is a result of cancer, nerve injury, or trauma; the pain, for example, that persists after a whiplash injury has healed is a secondary chronic pain syndrome; pain that persists after an episode of shingles is a secondary chronic neuropathic pain syndrome due to the virus having damaged nerves.

I sometimes run pain clinics on a Saturday when the hospital is quiet and I have more space to breathe. (This also enables me to get out of doing the Saturday cleaning at home.) I first met Helen on a Saturday. She was very fashionably dressed and was accompanied by her husband. They seemed like a loving and attentive couple; the partner who attends the appointment is sometimes too attentive to be helpful. She had the air of reticence I often see in patients who have been referred to a clinic named after a symptom. Neurosurgical clinic, cardiology clinic, respiratory clinic—these are clinics that inspire confidence of a cure. Not so a pain (management) clinic. Helen had been referred by a neurosurgeon because she had complained of pain in her arms and legs. The neurosurgeon had explained to her that there was no target for an operation on her spine that would cure her pain. She did not receive this information well. She'd had two previous operations on her neck and assumed that the symptoms she was experiencing were symptomatic of the same problem. So she could not understand why she was not being offered a third operation. (Not that the other two operations had resolved her symptoms. I am always amazed and a little bit envious of the faith that people have in surgeons and their scalpels.)

Helen was forty-eight and worked as an operating-room nurse, a job that involves standing for long periods of time, as an operation may last for three or four hours. Scrub nurses have to open up large, heavy trays and keep an eye on the number of swabs used in case one gets left inside the patient, a mistake that is easier to make than you think. Helen had two children in their teens and an older child who had children of his own. She was therefore mother, grandmother, wife, and NHS worker. I asked about her pain symptoms, and she said that she gets pain in most parts of her body on most days—a deep ache that gets worse as the day progresses. She experienced aching in her legs and her arms, felt constantly tired, could not remember things very well anymore, never woke up refreshed from sleep, and often had episodes of low mood for no real reason.

In my opinion, Helen was suffering from fibromyalgia syndrome, a poorly understood condition thought to be due to widespread dysfunction of the nervous system. We do not know what causes it, but patients complain of diffuse, achy, widespread tenderness throughout all four quadrants of the body. Environmental factors that can trigger fibromyalgia include soft tissue injuries such as those sustained in a road traffic accident as well as prolonged periods of psychological stress. The damaged muscles and ligaments from such an accident would normally result in acute pain lasting a few weeks, but in a genetically susceptible individual the stress (from both physical injury and psychological distress) can trigger the development of widespread and persistent pain.

Fibromyalgia and other, similar conditions such as chronic fatigue syndrome, which fall under the umbrella of "central sensitivity syndromes," may be triggered by certain types of infections, for example Lyme disease or viral hepatitis. Psychological stress, including from deployment to war, has been associated with the development of conditions such as fibromyalgia. There appears to be a mixture of genetic factors that renders an individual vulnerable to the development of central sensitivity syndromes. The genetic markers by themselves are not sufficient to cause these conditions, but interaction with environmental triggers can result in dysfunction of the neurological and hormonal system of the body, producing excitability of the pain system as well as related systems, such as those regulating sleep and mood.

Patients with fibromyalgia syndrome have associated sleep disturbance, chronic fatigue, and memory impairment. In addition, patients may present with widespread nonspecific and confusing (to myriad specialties) symptoms that can include bladder dysfunction without any obvious cause, breathlessness without respiratory disease, dizziness without neurological cause, burning, numbness, and tingling unrelated to any nerve injury, irritable bowel syndrome, and a host of other symptoms. The diagnosis of fibromyalgia is made after patients score the number of areas where they experience pain, together with the severity of their cognitive impairment, unrefreshing sleep, and fatigue. They also score any of the complaints they suffer with in order to arrive at a composite score. In the absence of any other condition that can better explain their symptoms, and if they meet the threshold score, the patient is given a diagnosis of fibromyalgia syndrome. Fibromyalgia may occur simultaneously with other chronic pain conditions, like osteoarthritis, rheumatoid arthritis, and systemic lupus erythematosus. Approximately 10 to 30 percent of patients with these rheumatic disorders also meet the criteria for a diagnosis of fibromyalgia syndrome.

When I explained this, Helen looked at me skeptically, as most patients do when I talk them through fibromyalgia. In a world where we believe in concrete, visible signs of disease and illness and where there is a scan or blood test for nearly everything, the idea that you can have dysfunctional nerves sounds like psychological gobbledygook to most people. I acknowledged that this was quite a lot of information to take in and that she was probably better off reading some of the leaflets we give to patients about the condition, pain management, and the role of clinical psychology in treating it. I offered her the option of coming back to see me if she did not understand the information; alternatively, if she had understood and wanted to embrace an approach that favored rehabilitation, then she could make an appointment to see one of our clinical psychologists. As first dates go in the pain clinic, it was not one of my more successful. I could sense her need for a more definitive solution to her problem.

A month later Helen phoned the pain clinic to ask for another appointment with me. When she and her husband sat down, she said to me, "I do not understand. I do not understand why if my arm is hurting it is not because of a trapped nerve in my neck. I do not understand how I can have pain in my arm that is not due to a trapped nerve or problem with my muscles. Is the pain psychological, and are you telling me I am imagining it?" She then started to cry. Her husband leaned toward her somewhat helplessly and gently stroked her thigh. When someone starts to cry in the clinic I often reflect on the expression "private pain" and the idea of being alone with your pain. I have never met a lonelier person than someone suffering with pain. Unlike a child that bumps its head and is soothed by the embrace of a parent, the chronic pain sufferer is not helped by the embrace of a loved one because the pain is unrelenting, and they cannot be held all day long.

The body has a system of nerves that form a network, and the dysfunction of these nerves results in the experience of persistent pain. I usually google "nervous system" and use the images section to talk a patient through what they are experiencing. I explain that the nervous system works a bit like a car alarm that goes off because somebody is breaking into the car; this is acute pain. When the intruders have been chased away, the alarm switches off. But sometimes the alarm goes off without there being an intruder, and in this case the problem is with the alarm itself. Both the pain and the loudness of the alarm are real, not imagined; the problem is not with the machinery but with the way the system functions. In the case of chronic pain, the alarm cannot be switched off once it is broken—the car owner can only learn to manage their response to the alarm.

This is the most distressing part of my job: offering answers with no hope of a cure. Being told that you have to manage a problem rather than having it cured—there is no disc that can be cut out, no bone that can be mended or joint that can be replaced—is difficult for most people to understand and accept.

I told Helen that our technology is not yet sufficiently advanced and that there is no scan or blood test that will prove she has chronic pain. She said her biggest problem was that when she tried to explain fibromyalgia syndrome to her colleagues at work they said, "Oh, that's a made-up condition, it's all in your head." They assumed she was trying to get out of work. Sadly, this kind of negative reaction is not uncommon, even among healthcare professionals, and is the result of the widespread lack of understanding why and how people develop chronic pain.

———————

Another primary chronic pain syndrome, complex regional pain syndrome (CRPS), deserves special mention. CRPS is a very unusual condition. It can develop spontaneously or as a result of trauma; for example, it can be related to having a fracture or some form of elective surgery. In either case it is characterized by pain that persists after the affected tissue has healed. The first reports of CRPS date back to the work of Silas Weir Mitchell, a US physician who described the condition in soldiers recovering from gunshot wounds in the American Civil War. The condition has been referred to as Sudeck's atrophy (referring to the thinning of bones), reflex sympathetic dystrophy (referring to the changes in color, temperature, and swelling of the affected area), and algodystrophy (from *algos*, meaning "pain," *dys*, meaning "bad," and *trophy*, meaning "nourishment," referring to the thinning of bones, blanching of skin, and generally sickly looking appearance of the affected limb); the various names reflect the theories physicians proposed to explain the condition or the predominant symptom observed. The term "complex regional pain syndrome" was coined in 1995 to resolve the confusion around nomenclature and facilitate further research.

The name given to the condition describes how little we understand about it: four words strung together to describe something that is still an enigma. When doctors use the term "complex" it means they don't understand what something is or why it occurs. The condition is "regional" in that it tends to affect an arm or a leg. "Pain" is a predominant symptom of the condition even though there isn't a reason for it to persist because the underlying injury is healed, and the pain is disproportionate to the inciting event. Doctors use the word "syndrome" when they are describing something whose origins they don't understand but that has particular changes and symptoms.

The best way to understand CRPS is to imagine that the usual response of the body to injury—which includes redness, swelling, pain, and sensitivity to touch in the injured region—becomes exaggerated and unregulated and persists for longer than is normal. The limb of a patient with CRPS may become hot or cold or change color. Patients may experience a tremor or inhibited movement in the limb that is unrelated to pain and exists because of a dysfunction in the nervous system that we cannot yet explain. They may have loss of hair in the affected region and changes to their skin, and their nails may stop growing as much or overgrow. The symptoms are often florid at the beginning and gradually reduce over time, but the pain persists in 15 percent of sufferers. Because the condition is so variable it is diagnosed by what are called the Budapest criteria—requiring these changes in color, swelling, and so on in sufficient prominence—so named because the consensus conference to determine the diagnostic criteria was held in Budapest in 2010.

The condition is difficult to treat and requires early rehabilitation. The person needs to have their pain reduced and needs to engage actively with a hand therapist for upper limb CRPS or a physiotherapist if it's the lower limb; spinal cord stimulation, an electrical treatment I will describe in chapter 8, can also help reduce the pain. Patients need to get the limb moving so that they can rewire the nerves and the immune system to dampen down the florid reaction. CRPS is a significant diagnosis when it occurs in the setting of an injury that has a compensation claim attached because the condition renders people considerably disabled, and the literature indicates that in the context of an ongoing medicolegal claim the prognosis for recovery is poor.

CRPS is challenging to diagnose, and experts tend to argue a great deal over whether an individual is truly suffering from it, particularly in medicolegal situations. When it is florid sometimes even the most ardent defendant-appointed pain expert cannot undermine the diagnosis, but for more subtle varieties medical experts may be hamstrung by the assessments of the clinicians who have seen the patient and have not utilized the Budapest criteria when making their assessment.

We understand that chronic pain is real pain, it is not imagined, and it is due to nerves that are not working properly. We know that patients with chronic pain do not feel normal and wake up each day with discomfort. And we know that disability and pain are not related in the chronic setting. If you break your arm there is a period of time when you cannot use it; until it heals you will not have normal function. With chronic pain, however, everything has healed, and so what determines your disability is your native resilience in the face of adversity, in the same way that what stops you from managing your diabetes appropriately is not access to healthcare or drugs or the way those drugs work but your approach to the management of the condition. If you manage it well and maintain your blood sugar within normal levels and choose to eat wisely, you can become prime minister or win four Olympic medals, but if you choose to live badly, the end result will be amputations, dialysis, and heart attacks. The difference is not the disease but the way you approach and manage your condition.

It is the fact that people have this choice that sometimes makes it difficult for me in the clinic. I love my father. My father is a GP and has been for forty-odd years and still works. I would love my father to live forever. I just had a conversation with him after he went to see his cardiologist, who did the usual stress test and then prescribed yet another drug to lower his cholesterol. However, my dad, who lives in South Africa, loves fries from the local roadhouse and a spicy, high-fat, salty snack called Bombay mix. He will not change his lifestyle, so the only thing that will help him live well and longer is to stop eating Bombay mix and fries. When we love people it is so difficult to deal with the poor choices they make. But when I offer people in the pain clinic the opportunity to rehabilitate, I am often not privy to their narrative. What role does pain and the disability from pain play in their overall life? How much do they need their disability check as a consequence of their fibromyalgia? What role does their chronic lower-back pain play in maintaining their relationship with their spouse? If they got better, would their partner still love them? What is the family dynamic—do they rule the roost with their chronic pain as the center of attention? While it is the right thing to do from an evidence-based point of view, offering patients rehabilitation often completely unsettles families.

There are three ways to deal with chronic pain. You can spend your time agitating about the alarm going off; you can try to ignore it (but ignoring it actually takes up more energy); or you can learn to live with this deeply unpleasant situation. There will be times in your life when the alarm is louder, particularly when you are stressed or upset about something, but there will be times when you busy yourself with other activities and the alarm makes very little difference to your life. Having said this, unlike the management of other chronic conditions such as diabetes or high blood pressure, which you do not feel unless you become acutely unwell, pain is something that is there with you all the time. It demands your attention, like a small dog yapping at your heels. It distracts, frustrates, and distresses people and affects their identity; if they are constantly being harassed by pain they cannot accomplish tasks and lay down the memories that make them who they are.

It is not entirely clear to medical science why some people who have an injury or a physical problem go on to develop pain that persists, but we think there are underlying biological reasons. Patients with chronic pain may have different enzymes that predispose them to developing persistent pain. There is an enzyme that breaks down the neuro-transmitters noradrenaline and serotonin, chemicals involved in both mood and pain, and people who develop chronic pain have been found to have abnormal versions of these enzymes; they end up with lower levels of these pain-relieving and mood-uplifting chemicals. We also think that psychological factors (thoughts and feelings) can cause physi-ological changes, particularly changes in brain chemistry, which might lead to a person's developing chronic pain. In a patient prone to catastrophizing—magnifying, ruminating, and feeling helpless in the face of adversity—this can result in a change in the normal ecosystem of brain connections (the connectome), which then affects the way the muscles and nerves function; instead of the pain alarm switching off after the body has healed, it continues to blare. Genetic factors can also contribute to changes in the way the harm-sensing receptors function; in some individuals they are more abundant and more sensitive, leading to a greater propensity to neural inflammation or prolonged injury. There can be changes within the connections in the spinal cord, leading to an increase in the number of excitatory chemicals released at the dorsal horn in response to a signal from receptors in the skin and other tissues, causing the spinal cord to become "wound up" and send more messages to the brain.

The brain itself changes when an individual has developed chronic pain and suffered with it for a long time. The amount of gray matter in the brain of a patient with chronic back pain reduces over time and new connections are formed, which makes the brain more alert and respon-sive to the experience of pain. It is a bitter irony that the individuals who develop chronic pain become more attuned to feeling pain because of the changes, triggered by pain, which develop in the nervous system that make it even more sensitive. A chronic pain sufferer presenting for surgery that is unrelated to their chronic pain is therefore more likely to experience severe postoperative pain due to this existing dysfunction.

The idea of "pain stickiness" has been proposed to try to understand why some people develop chronic pain after an injury and others do not: pain sticks because of a host of genetic, environmental, social, and psychological factors. We observe physiological changes in individuals suffering chronic pain, such as alterations in nervous system connections in the brain, but we do not yet understand what it means. Ultimately, we yearn for a biobehavioral model to explain why chronic pain develops.

In patients who suffer with catastrophizing thoughts, brain areas are activated that cause the pain pathways to be amplified. When you amplify a pain pathway it is like new channels being formed within a riverbed. Initially there is a trickle of water, but as more and more water is released the furrows become wider and wider and eventually the trickle becomes a raging torrent. This can be in part due to behavioral changes that a person makes when they experience pain. Patients who have ruptured their Achilles tendon and are still experiencing pain may think that it has not healed properly and may stop walking. Then the tendon may thicken and become immobile. So when they do try to move they experience pain, which reinforces the idea that something is wrong and that the doctors have not fixed it properly. They may see a doctor who does not encourage them to rehabilitate and tells them the pain should be their guide, whereupon the fact that they continue to experience pain results in their moving even less, aggravating the situation. We now encourage people to rehabilitate in the presence of pain and try to reassure them that theirs is a safe pain. But not every doctor embraces this paradigm.

My heart sinks when I come across a patient who says they last worked when Backstreet Boys first got together because I know that they have a lifestyle that is ruled by chronic pain. Breaking somebody like this out of the cycle of chronic pain is almost impossible because I am not just trying to manage a clinical problem, I am also threatening to dismantle somebody's world. As a taxpayer and clinician I find myself thinking, "This patient could be rehabilitated if they only wanted to be," but then I consider the extent to which our society facilitates chronic pain in so many ways. Persistent pain affects identity formation, which relies on engaging in day-to-day activities. If society will continue to support people who adopt a disabled role, then it is likely that some people will take this path. As mentioned, chronic pain is not an indicator of tissue damage, and so there are individuals who learn to function despite being accompanied by daily pain, and then there are others who choose to withdraw and are disabled by their pain. This is not always a conscious choice, however.

In my ten years as a pain consultant it has become increasingly obvious to me that much of what we do in modern medicine is fairly futile—unless we are treating cancers that we can cut out and perhaps prolong life or are treating trauma or infectious disease. In everything else we are at the wrong point of the thrust of treatment. We should not be the tip of the spear; we should be somewhere on the side, and the tip of the spear should be those clinicians who can facilitate and motivate change in the individual and on a wider social scale. When you live in a world that tells you to eat less sugar but only two aisles in the supermarket actually contain anything nutritious, you are dealing with a society that is against getting people to live well.

The healthcare system, with its doctors, nurses, and allied health-care professionals, hospitals, and treatment centers, is constructed on the foundation of being able to arrive at a single diagnosis for a set of complaints and to decree management based on that diagnosis. Health-care is an industry (as it has been since the time of ancient Egypt), and sometimes the maintenance of the industry's status quo becomes a more powerful motivator for the people who work in that industry than alleviating the suffering of the people the industry was designed to serve. It becomes difficult therefore for the industry to acknowledge its limitations when faced with someone with chronic pain, so rather than attempting a different, nonmedicated approach, we pass along patients from station to station like passengers on a train with no specific destination.

As we have seen in this chapter, chronic pain is incredibly hard to accept for both doctors and patients because the pain is not linked to a disease or injury that can easily be diagnosed with our current technology. There is no blood test or scan that can determine chronic pain. Given these challenges in even diagnosing a chronic pain condition, we will now look at the ways that various medical professions have tried to manage, and even "cure," chronic pain.

CHAPTER SIX

Pain Management: Needles, Narcotics, and Knives

Whenever I am asked to teach a group of medical students, junior doctors, or even a group of my peers, I often quote the words of the Roman emperor Marcus Aurelius (121–180 CE). He was fond of reminding himself and others, "Everything we hear is an opinion, not a fact, and everything we see is a perspective, not the truth." Marcus Aurelius was schooled in the philosophy of Stoicism, and his *Meditations*, a book recording his private thoughts and a reminder to himself to be humble, patient, empathetic, generous, and resolute in the face of adversity, is still widely read today. It is said that he employed somebody to walk beside him whose role was to whisper in his ear, "You are just a man. You are just a man."

When it comes to discussing our approach and methods for easing suffering—in medicine generally and in the fledgling field of pain medicine particularly—I find it helpful to remind myself of Marcus Aurelius's words: that my perspective is but a fleeting moment in time, and that a century from today the procedures that we argue, fuss, and fight about will be viewed with a mixture of amusement, curiosity, and amazement. Our patterns of thinking, our scientific limitations and hubris will be reflected on in the same way that we now shake our heads at the physicians who employed bloodletting as a medical panacea, back when illness was believed to be due to an imbalance of the four humors.

The practice of engaging in a trial-and-error approach to therapeutic interventions, with no clear idea of the pathology being treated, is not new in clinical medicine. Our incomplete understanding of disease can lead to ill-conceived interventions that, today as in centuries past, can be responsible for a great deal of harm. Oliver Wendell Holmes, a US Supreme Court Justice in the early twentieth century, is quoted as saying, "From making the cure of illness more insufferable than the illness itself, Lord deliver us." It is perhaps also why physicians from antiquity felt the need to state explicitly *Primum non nocere*—First, do no harm.

Harm is an interesting concept—it is not always the obvious harm of causing death or paralysis by a misplaced needle, a poorly chosen incision, or a carelessly used drug. Harm can be facilitating a patient who uses our repeat injection procedure to justify their disability to the Department of Labor. Harm can be that their reliance on our needle causes them not to engage in more effective behaviors such as exercise or mindfulness to manage their pain and disability. Harm can be that we slowly cause their spine to become osteoporotic with repeated steroid injections, but so slowly that our injections are not seen as the cause.

Modern interventional techniques to manage pain can, when mishandled, cause harm. These interventions can be broadly broken down into three categories: those that cause destruction of nerves; those in which therapeutic substances such as local anesthetics and steroids are injected around joints or nerves; and those that use electricity to stimulate the nervous system (spinal cord stimulation). These techniques fall naturally within the skill set of anesthesiologists, who are involved in blocking nerves in the perioperative setting to facilitate surgery and on obstetrics wards to relieve the pain of childbirth. Patients with nonspecific lower-back pain and patients with peripheral nerve pain (sciatica) due to irritation of spinal nerve roots are therefore easily treated with injection therapies by the same clinicians. There is cross-over between using a needle and syringe in anesthesia and the techniques of pain medicine; however, in the former the procedure is done to ensure pain relief for only a short while during and after surgery, whereas in the latter we hope to manage pain over the long term.

Modern interventional techniques to manage pain—now slowly being decommissioned in many parts of the world due to a lack of evidence for them—have relied heavily on the use of steroids. Steroids were first discovered in the 1920s, but it was only after the Second World War that patients with rheumatoid arthritis were successfully treated with the steroid cortisone. The spectacular response to steroids in patients with rheumatoid arthritis has made their use fundamental to the treatment of inflammatory rheumatological conditions. The powerful inflammatory suppressive nature of steroids led to the application of the substances to what were regarded as inflamed areas of the spine, particularly nerve roots. Initial studies applying steroids to inflamed nerve roots were carried out on series of patients, without the randomization of patients with specific diagnoses into treatment and nontreatment groups in order to assess if it was the treatment making the difference or if the result was down to chance. Between 1953 and 1993 most studies were retrospective, looking back at patients who had been treated with epidural steroids and then quoting the results. Retrospective studies are famously biased, potentially due to a lack of negative outcomes being recorded. The first prospective studies yielded mixed results: some showed a striking benefit from epidural steroid injection, while others showed the injection of steroids to be no more effective than a placebo injection in relieving chronic symptoms caused by proven lumbar disc herniation. Despite this very weak evidence, the practice of epidural steroid injection into the spine continues today.

There was a time before the Nuremberg Code (1947) and the Declaration of Helsinki (1964) and the advent of evidence-based medicine when doctors experimented freely without having to conduct ethically approved, carefully constructed studies, overseen by committees and aimed at identifying the treatment effect of an intervention (the actual effect minus the placebo effect) and evaluating its safety. Lumbar facet joints, which are the sliding joints between the bones of the spine, were first described as a possible source of back pain in 1911, and the term "facet joint syndrome" was coined in 1933. In order to learn whether back pain was originating in these joints a so-called provocative study was undertaken (as opposed to a therapeutic study) by a team of doctors led by Carl Hirsch in 1963. His team injected less than 0.3 ml saltwater, which is a highly irritating substance, into the facet joints of volunteers using X-rays to ensure that the needle was in the correct position. Pain was felt in the lower back, sacroiliac, and gluteal areas of the patients who were injected. The researchers described the results in their paper "Studies on Pain Following Injections of Hypertonic Saline in Discs, Joints and Ligaments in the Low Back," stating: "Pain occurred after a few seconds and was very annoying."

The first recorded therapeutic injection of the facet joints with steroid and local anesthetics was done by two orthopedic surgeons, Vert Mooney and James Robertson, in 1976, followed by a radiologist, I. W. McCall, and his team in 1979. This therapy was in increasingly widespread use for lower-back pain over the next thirty years, until it was found to be without any benefit for nonspecific back pain.

There is controversy surrounding who the first person was to inject substances into the epidural space to treat patients with pain; credit is given to the French neurologist Jean-Athanase Sicard (1872–1929); however, around the same time the physician Fernand Cathelin (1873–1960), also from Paris, had been treating patients with epidural injections for several months. The first series of nine patients who underwent epidural injections included one suffering with syphilis that had invaded the nervous system, four patients suffering with lower-back pain, and four with pain due to nerve injury. Sicard, who performed these procedures, became famous as a pain doctor during the First World War, when he used alcohol to destroy peripheral nerves in order to alleviate pain, particularly in conditions such as complex regional pain syndrome. Perhaps it was because he published his ideas in French that they were not widely read and using the epidural route to help with pain did not spread as quickly as other medical innovations. We would refer to this phenomenon today as publication bias, a problem that continues to thwart medical progress. Curiously, the disease we most commonly use epidural steroid injections for today (aside from epidurals used for pain relief after surgery and during childbirth), which is herniated discs causing compression or irritation of nerves, commonly referred to as sciatica, was not widely known until 1934. It is quite possible therefore that Sicard's patients happened to be suffering with sciatica and their serendipitous response promoted the therapy.

Controversy also exists around whether the American neurologist James Corning or the German surgeon August Bier was the first to perform a spinal anesthetic. Spinal anesthetics, or subarachnoid blocks, differ from epidural injections in that the epidural space is separated from the subarachnoid space by the dura (hence epi-dural space). The cerebro-spinal fluid flows in the subarachnoid space, and anything injected there has direct access to the nerves of the spinal cord rather than just the nerves *exiting* the spinal cord, as in the case of an epidural injection. In 1885 Corning injected cocaine into the intrathecal space (around the spinal cord) of a man who was diagnosed as suffering with habituated masturbation and seminal incontinence; the long-term outcome of the treatment is not known. In 1898 Bier injected 15 milligrams of cocaine into the fluid surrounding the spinal cord and successfully anesthetized one of his assistants, August Hildebrandt, having previously used the technique on a series of six patients. He tested the effect of the local anesthetic by applying a burning cigar to Hildebrandt's skin, a strong blow to the shin with an iron hammer, and strong pressure and traction on one of his testicles. We tend not to use these methods today, preferring the application of ethyl chloride, which feels cold when it evaporates from the skin; patients who have had a local anesthetic introduced into the cerebral spinal fluid feel the sensation as warm or not at all.

Local anesthetic introduced to the intrathecal or subarachnoid space is used for operations performed below the belly button, including cesarean sections, but this space is limited in terms of its usefulness regarding chronic pain. Intrathecal drug delivery using very low dose local anesthetic, opioids, or conotoxin (a neurotoxin derived from the sea snail *Conus magus*) to block the calcium channels in nerves can be used to treat chronic pain, using an implanted pump that is periodically refilled. The helpfulness of this therapy, however, is limited, and complications associated with the technique have made it unpopular. For cancer pain, alcohol or phenol is sometimes used to destroy nerves using the intrathecal route but only when life expectancy is less than the time that it would take for nerves to start regenerating (about six months), which would cause more severe pain.

The evolution of interventional pain therapies blossomed from the therapeutic tradition of offering patients with intractable cancer pain relief from their suffering through destructive neurosurgical and anesthetic procedures, prior to the gradual development of more effective oncology therapies and holistic palliative care services in the latter part of the twentieth century. These destructive therapies were acceptable because patients often did not live very long, so the long-term complications from destroying nerves were often not experienced—the patient would succumb before manifesting the pain from regenerating nerves or the ignominy of incontinence. Some nerve-destruction techniques are still practiced today, such as cordotomies, first performed in 1912, in which the pain pathways at the level of the cervical spinal cord are destroyed in patients who have pain on one side of their body. This is an invasive therapy with potentially severe complications, but it does offer individuals with relatively short life expectancy relief from pain and suffering. Similarly the nerve supply to the pancreas can be destroyed by injecting alcohol, phenol, or glycerol, disrupting the normal architecture of the nerves by dissolving protective fat layers and obliterating the orderly sodium and potassium channels on the surface of the nerve, which enable electricity to flow. But unlike distinct fiber optic cables transmitting discrete packets of information, nerves are intermingled and inseparable, and the destruction of nerves that transmit pain also disrupts nerves that control the function of the gut, causing diarrhea, bladder and bowel incontinence, and paralysis of the legs.

These therapies are now used more infrequently and judiciously because cancer survival rates are greater, and we rely much more on painkillers and anticancer drugs to manage cancer pain. Because nerves regenerate when you destroy them with glycerol and alcohol, and the regenerating nerve expresses new sodium and potassium channels in a haphazard fashion when they do regenerate, the long-term consequences of nerve destruction are worse than the pain they were originally designed to treat. The new channels act like the frayed end of a ruptured electrical cable, sparking, burning, and fizzing as they desperately try to conduct electrically encoded information to the section of nerve they have been separated from. This is called "deafferentation pain." Techniques using alcohol, phenol, or glycerol to destroy nerves are therefore not appropriate for patients who suffer with chronic nonmalignant (cancer) pain. Because such patients are likely to live for a long time these techniques will result in worse neuropathic pain in the years following the procedure.

Interventional techniques using needles to reach nerve bundles evolved over time to be incorporated into the specialty of anesthesiology. As medical specialties became more refined, palliative medicine took over the symptom management of cancer sufferers and anesthesiologists working in pain medicine migrated toward treating patients with nonmalignant pain. They too employed steroids and local anesthetics but without destroying nerves, thus circumventing the issue of deafferentation pain. It is interesting to note that the techniques and application of novel therapeutic substances preceded any theory of why the person had pain, rather than the other way around. Albert Einstein was particularly critical of any approach where technique evolved before theoretical explanation, and Sir Arthur Conan Doyle, himself a physician, expressed through his fictional character Sherlock Holmes his suspicion of the approach of theorizing before one possesses data, as this would, in his view, lead to twisting facts to suit theories rather than using facts to develop theories.

Throughout the twentieth century anesthesiologists involved in treating chronic pain focused on pain relief rather than curing the underlying condition, which (most) clinicians soon realized was not possible. Many pain clinics largely ignored or were ill-equipped or unmotivated to incorporate behavioral and physiotherapy treatments to manage the loss of function patients with persistent pain suffered, as well as the psychological distress that followed from this loss of the ability to work and perform activities of daily living. Millions of dollars have been spent on the injection of local anesthetic and steroids into lumbar facet joints as well as epidural steroid injections over the past fifty years, despite a very poor theoretical or evidence-based platform for their usefulness and their questionable value to patients. This century has seen a shift to more managed healthcare with greater involvement of the organizations, both public and private, that pay for treatments and a rigorous assessment of the cost versus benefit of treatments and the scientific evidence that underpins the therapy.

While interventional pain medicine doctors with an anesthetic background have been happily flying solo, injecting patients with local anesthetics and steroids, burning nerves, and prescribing drugs, there has been a parallel evolution in the understanding of chronic pain as a disease rather than merely a symptom. Here the chronification of pain is understood to be due to a complex dysfunction of the nervous system. There has been and continues to be a surreal disregard by some doctors of the complexity of chronic pain and the influence of psychological and social factors on its maintenance and evolution and the myriad pathways to and from the spinal cord that influence the experience of pain—be it acute pain due to injury or chronic pain, either cancerous or noncancerous. Anesthesiologists who practice in the field of pain medicine and perform interventions are often labeled "needle jockeys" by psychologists and pain management physiotherapists, who marvel at the obliviousness of the individual wielding the needle to the psychological distress of the patient, and at the limited literature on the safety, efficacy, fiscal neutrality, and effectiveness of interventional pain techniques.

There is a great deal of expressed and hidden anger on the part of psychologists and pain management physiotherapists, as well as occupational therapists and nurses who work in pain medicine, with regard to the amount of money that has been poured into interventional techniques. This does lead to very interesting (and infrequent) meetings of multidisciplinary pain societies, and I sometimes wonder if this ability by most pain anesthesiologists to ignore the widely available literature on the complexity of chronic pain is deliberate or merely stems from the absolute belief in their own ability to relieve pain. Are they continuing to provide these therapies because they only ever see their benefit at the six-week mark, when the patient is reviewed in their clinic? Are they perfectly happy to accept a therapy with only a short-term benefit even in the absence of collecting data on possible long-term side effects? The ongoing provision of injection therapies to patients with chronic non-cancer pain and the patient's descent into a vortex of disability is perhaps best illustrated with Diane's story.

Diane had a difficult and troubled childhood and responded with displays of anger and anxiety to the conflict between her parents. Her father divorced her mother when Diane was sixteen, and the family was forced to move far away from Diane's circle of friends and social support structures. When she was in her twenties, she suffered with frequent headaches and a strain to her lower back. She also had abdominal pain that her GP could never pin down to a specific disease, and she complained bitterly about painful periods.

Her lower back started to hurt during her first pregnancy, and as the months progressed her backache became worse. Her pain resolved after the birth of her child but returned when she had her second child. In her early thirties the back pain became continuous. She found her job as a cleaner unrewarding, and the management were unsympathetic toward her pain and the limitations it placed on her ability to work. She sought help from her GP, who prescribed codeine and paracetamol to help her symptoms. He was unable to provide her with an explanation of why she had lower-back pain, dismissing it as a common occurrence in people who did manual work, but he performed an examination to rule out anything dangerous in the form of a tumor or a fracture of her spine. She had full control of her bladder and bowel functions, so he was not worried that she had a disc prolapse compressing the nerves radiating from her spinal cord. The persistence of her back pain and her associated distress led to a referral to an orthopedic clinic, where she was seen by a nurse. The nurse explained that the pain Diane was experiencing was coming from her sacroiliac joint, which is an immovable joint between the hips and the sacrum (a triangular bone in the lower back).

Diane continued to complain of back pain and increasingly found that it was interfering with her ability to work and disrupting her sleep. She felt generally unwell—tired, achy, lethargic, and depressed. By the time she was in her late thirties her pain had become so continuous and limiting that she stopped working. Her GP continued to treat her symptoms with increasingly strong medication, and eventually she was started on morphine.

Diane was eventually referred to a pain clinic, where she met an anesthesiologist with an interest in pain medicine, who worked in a clinic without a psychologist or physiotherapist. After prodding her cursorily for five minutes, he concluded that she had pain emanating from the sacroiliac joint, into which he injected local anesthetic and steroid as an outpatient procedure, using an X-ray machine to guide the needle into the joint, which sounds wonderfully technical and not surprisingly has a tremendous placebo effect.

Sadly but not unpredictably, the injection into Diane's lower back and into the sacroiliac joint gave her only five weeks of pain relief. But the pain consultant considered this to be a positive outcome and scheduled her for radiofrequency denervation of the nerve supplying the joint, which means he planned to insert a needle into the nerve and apply a radiofrequency wave to excite the ions within the nerve, destroying it on a semipermanent basis. Diane continued to seek answers from her GP when her pain continued despite these interventions, but he still could not furnish her with an explanation. When the sacroiliac joint injection did not improve her symptoms, her pain consultant moved on to other joints within the lumbar region of her spine, performing first local anesthetic and steroid injections into the facet joints, followed by procedures to burn the nerve giving sensation to these joints. He then performed an epidural injection, bathing the nerves that leave the spine in a mixture of local anesthetic and steroid.

Diane's injection journey continued for a year, at the end of which she was receiving disability and no longer able to do her own housework. When the pain consultant had run out of sites to inject, he referred her to physiotherapy for an assessment and finally discharged her to a community pain management clinic that was run by a psychologist and a physiotherapist.

Diane's story follows a common theme. Episodes of acute non-specific lower-back pain beginning in pregnancy and transforming into persistent pain are not unusual. We think that these episodes of acute pain hammer away at the pain alarm system, eventually rendering it permanently active by changing connections in the brain and spinal cord. Diane had two psychological factors, anxiety and depression, making her vulnerable to the development of chronic pain. In addition, environmental factors slowly impacted her nervous system, damaging it with an unfavorable chemical environment; for example, her cigarette smoking nagged at cells in the lungs, causing bouts of inflammation and infection and would eventually lead to cancerous transformation or persistent breathlessness. The clinic she was referred to applied treatments in line with those practiced by pain clinics across the world, both public and private. Diane said that these consultations never lasted long and always ended with her either receiving the promise of a cure from an injection or being referred back to her GP for further medication management.

An intervention may be considered successful if it simply reduces pain, regardless of whether that pain reduction results in improved function and psychological well-being. In contrast, the pain center I work at is very active in applying the growing body of evidence with regard to the biological, psychological, and social factors influencing persistent pain conditions. Still, when I go to meetings with clinicians committed to cure at the end of a needle, I sometimes feel that I am sitting alone outside an interventional pain medicine town when it comes to the way that I practice pain medicine.

In October 2018 I attended a meeting of interventional pain doctors considered experts in their field. We were in a chateau in Fontenay-Trésigny in the heart of the Brie region in France, and the meeting was organized and hosted by a company that manufactures the equipment used for interventional pain techniques, including the machine used to ablate nerves—one of the techniques tried, unsuccessfully, on Diane. Injection of local anesthetic and steroid around facet joints gives the patient only approximately three to six months of pain relief at best, and so the idea evolved of using radio waves to burn the nerve that supplies sensation to these joints, making the pain relief more permanent—this is nerve ablation. This nerve runs over the transverse process of a vertebral body. Imagine a narrow stream running over a craggy rock in the middle of a mountain, and now imagine you have to access this target with a fine needle inserted from beyond the stratosphere; that might give you some idea of the complexity of the technique. It is very easy to miss the water but quite easy to hit a structure adjacent to the water, which may result in significant complications.

The assembled experts were discussing how to insert the needle and whether using X-ray or ultrasound to guide the needle was more appropriate. The amount of heat generated by the needle tip as well as whether to use a pulsed or continuous radiofrequency were discussed at length. I suggested that as we did not even understand the nature of the disease we were trying to treat, nor its pathological or anatomical basis, discussions about the techniques of specific therapies were premature. In fact such discussions were ludicrous given what we do know about the short-term benefit of these therapies, the lack of long-term benefit, and the potential for complications such as nerve damage.

They agreed grudgingly with my logic and reluctantly admitted that we have got ourselves into a real fix as pain experts. There is no doubt that some patients do derive relief from interventional pain therapies such as radiofrequency ablation of the nerve that supplies sensation to the facet joints. There are also patients who benefit from local anesthetic and steroid injections, if only on a temporary basis to facilitate rehabilitation—enabling them to walk a little farther, make their bed, and wash the dishes. But we have failed to collect evidence concerning which patients benefit from this therapy. We don't understand the precise mechanism of why these therapies should work in a chronic pain condition where pain persists due to changes in the way the nerves function. Is it the placebo effect, whereby the patient believes they will experience pain relief, and chemicals are therefore released in the brain that block pain signals? Does our meddling perhaps activate the patient's native descending inhibitory pain pathways? Or is it something that happens on a cellular level?

We have failed as a specialty to identify the exact nature of the pain that is responsive to these injections. As a result, the evidence broadly points to the injections being of very low utility in the management of nonspecific lower-back pain. Those responsible for paying for healthcare (governments and private health insurers) are therefore refusing to fund these treatments. The result is that the patients who *do* benefit from these procedures, usually the elderly, are now being excluded, along with those patients for whom the therapy was inappropriate.

———

Our ability to design effective treatments for chronic pain is thwarted by our infantile grasp of the complexities of the pain alarm system.

What is more feasible and important at this stage is attempting to help the sufferer alter their behavioral response to the pain alarm. To take a common example, when it comes to lower-back pain, we do not have the drugs or the interventional pain therapies that can permanently modulate the pain. We have made some progress with modulating pain from damaged nerves in the form of spinal cord stimulators, but the management of nonspecific lower-back pain remains elusive. In the meantime, this condition places an enormous burden on society, leading to massive losses in economic activity. While we cannot cure it, we can help people understand that these abnormal sensations are not dangerous and teach them to successfully live with persistent pain. Regrettably, though, we have not designed healthcare with this ethos in mind.

What we have done is create pain clinics consisting largely of doctors drawn from an anesthetic background who attempt to make a diagnosis based on a philosophy that aims to attribute a complex condition to a problem within a single structure, such as a facet joint; this philosophy is rooted in the old Cartesian model of pain, which you may remember from chapter 2. This approach leads to inferring a relationship between changes on a patient's MRI scan and the pain that they experience, despite the fact that the visible changes will in fact occur in as many people untroubled by pain as they do in people with persistent pain. If you scan a thirty-year-old who has no back pain, the MRI will show all the same changes associated with somebody who has back pain. There is therefore little basis for making a pathological connection between a scan and an individual's pain.

As a specialty we have continued to perform these interventionalist procedures and amass the fees for them, creating an industry with treatments that we can market and sell. At the meeting in Fontenay-Trésigny I was told by one of the company representatives that nobody could understand why doctors from a certain country could perform so many procedures during a session. They then found out that these doctors would bring the patient back on three occasions rather than doing the full procedure at one sitting. The reason for this apparent productivity was purely to collect triple the fee.

The National Institute of Care Excellence (NICE) in the UK has now issued guidance on the management of lower-back pain, curtailing the use of interventional pain therapies based on the available evidence. NICE is the government body that evaluates whether therapies used in the National Health Service have an evidence base and guides clinicians on service implementation. These agreements regarding interventional pain therapies were reached with great difficulty due to the vast vested interests at play. While in the UK NICE has taken charge, through commissioning groups, of what services can be provided, the rest of the world is still very much like the Wild West, where a doctor can take a course over a weekend and be burning nerves in someone's back on Monday. Because doctors make a living from performing procedures, these continue to be offered despite little evidence of their efficacy and an incomplete understanding of the condition for which they are used. This goes against the age-old advice with which we started this chapter: *Primum non nocere*—First, do no harm.

Many pain physicians reading this will be quite irritated and annoyed by it. I have done all the procedures I have discussed, and I have followed up patients at length and do appreciate that in some individuals these methods may well reduce chronic nonmalignant pain. I know, however, that in the long term these interventions do not improve function, nor do they reduce the distress associated with not understanding why the pain has persisted. In my view, when dealing with younger patients repeated interventional techniques lead to collusion between the pain physician and the sufferer of chronic pain by validating the condition and facilitating disability. Patients like Diane are left no better off following multiple interventions over a year or more. In fact, Diane was left more disabled, more distressed, with less understanding of her chronic pain, and caught in a spiral of worsening distress and disability. All I have tried to do in this chapter is to highlight the fact that, in the face of such distress, it is quite easy for doctors to offer a therapy in the absence of robust outcome measures.

Doctors in all parts of the world are at the tip of the spear when it comes to the management of chronic noncancer pain. In my opinion this is a fundamental flaw in the system, because reducing the distress and disability associated with chronic pain—which is essentially a condition that cannot be cured—does not rely on the skills of doctors. This is not a broken bone that needs to be plated and fixed, where knowledge about the anatomy of the arm and the correct use of a plate and screws is essential. It is not a condition for which we have appropriate and accurate tests to diagnose the degree of pathology and prescribe medications to manage the disturbance. Chronic pain management is not diabetes, which consists of assessing the degree of blood sugar control that an individual has and carefully titrating insulin to produce a blood sugar within normal range. This type of disease management is what doctors are good at; that is why we learn physiology, pharmacology, and anatomy in medical school. Even in a condition as clear-cut as diabetes, once the doctor has done their initial job the role of psychologists and nurses is in fact more important because maintaining compliance is much more impactful in the long term than any changes in medication.

Pain medicine is a fascinating example of a specialty that has never come to grips with the nature of the condition it seeks to cure, but which has plowed ahead with the application of therapies borrowed from other specialties in order to justify its existence. Anesthesiologists involved in pain medicine have piggybacked on various revolutions, including the fashion of treating acute pain aggressively with opioid therapy and injecting steroids into the spine to treat chronic noncancer pain.

Interventional pain medicine is a perfect example of how an industry can be built around poor science and dubious data. As a branch of medicine it has evolved from incomplete models and entrenched beliefs related to a complex phenomenon. The rush to action in the face of suffering has come before careful consideration of the evidence; therapies have been accepted without critical evaluation, and because they do not appear to cause obvious harm their practice has continued. A more complete approach is needed, an intersecting Venn diagram of needles, knives, analgesics, and interdisciplinary rehabilitation aimed at behavioral change, with the intersection of the diagram being a reduction in the distress and disability associated with persistent pain.

CHAPTER SEVEN

A Journey of a Thousand Miles . . .

Patients who present with intractable chronic pain do not fit a neat diagnostic label and are like people who arrive by train at a frontier town in the old Wild West. The pain clinic is often their "last chance saloon" on a journey that has passed through many cities and many different specialties, several of which we explored in the previous chapter. Patients have often been subjected to numerous treatments and investigations on the way to the pain clinic and arrive confused, wary, weary, and desperate, like survivors of an apocalypse or entrepreneurs who have lost fortunes and optimism at the different stations on their journey to cure and understand their chronic pain. Patients with back pain, for example, may have been reviewed by neurosurgeons or spinal orthopedic surgeons, with no explanation provided for their ongoing symptoms. The time spent at these stations has often resulted in increased distress, as they are promised therapies that ultimately do not relieve their symptoms or, even worse, make them feel that they are among the incurable and must live with the knowledge that their condition will continue to progress, that they are cursed with ever-increasing disability. Many come to believe that a life spent in a wheelchair is their only future.

On their journey they may have encountered pain physicians who believe that the sufferer of chronic pain can be helped with a single intervention. These are the mavericks of the frontier town promising a magic bullet—one injection or tablet that will resolve the unpleasantness the person is experiencing and restore them to fully functioning members of society. Then there are pain physicians living on the other side of the tracks, of whom I am one, who believe that the problems the chronic pain patient suffers are more complex and require greater input from a wider variety of people. These are the reformed gunslingers who now work collaboratively with psychologists and physiotherapists to reduce the distress and disability associated with persistent pain, understanding that gunslingers miss more often than they hit, that sometimes the bullet causes unintended harm, and that violence is not the answer. Hope and desperation are the baggage accompanying patients with persistent pain as they disembark from the train, and this makes them vulnerable to promises of a therapy involving a single intervention that will facilitate a cure.

One of the phrases that patients with persistent pain often use and that makes my heart sink whenever I hear it is "I will try anything." This statement implies the patient's desperate passivity ("Do something to me, Doctor. Fix me.") and a lack of active engagement in their management and care. Patients get to a point where they will cling to any fragment of poorly constructed hope or wild explanation of the cause of their pain, due to desperation and a belief in the power of the doctor to mend their broken bodies. We doctors often collude with patients by not emphasizing the changes in behavior they need to make to improve their health. The equivalent example is going to the gym but not changing your diet to achieve the benefits you seek.

For many patients a pain clinic is their final and often reluctant destination. The route to the clinic is often meandering, the patients moving between orthopedic and spinal surgeons, with detours to neuro-surgeons and often to pain clinics that have offered only interventions, not rehabilitation. This journey adds to the distress of these individuals. Some arrive in weary resignation and despair and are perhaps more easily inspired to change and be educated about chronic pain. Others arrive angry and spoiling for a fight because of the failure of doctors to restore them to normal health; they are often caught in a vortex of disability from which they do not feel they can escape.

Pain and suffering are intimately related. Patients who experience persistent unpleasant body sensations are often fearful, anxious, and depressed. Pain behavior is another component of the concept of pain and refers to those things that people say and do when they are suffering. These behaviors are a way for individuals to communicate their pain and distress. If we want to understand somebody's perception of pain and their response to these abnormal sensations, it is essential to consider their experience as a whole, to examine the biological changes potentially underpinning the pain, such as inflammation, along with any psychological factors and the society and cultural environment within which they live. If we focus on only one of these aspects—biological, psychological, or social—our formulation of the problem the person is suffering with will be inadequate, and any attempts to manage their chronic pain and associated distress and disability will inevitably result in failure.

The preceding chapter discussed the inadequacies of trying to manage chronic pain as if the problem was purely biological, without any acknowledgment of the psychosocial factors involved—finding a target and obliterating it with a needle-based intervention. This chapter outlines an approach that integrates all facets of pain. The aim of pain management programs is to enable people with chronic pain to achieve as normal a life as possible by reducing physical disability and emotional distress.

———————

Patients who come to my pain clinic can be grouped into three broad categories based on the biological mechanism of their pain. First, there are patients who suffer with pain due to disease or injury of a nerve (neuropathic pain). Of these, the most common group is patients who have had a prolapse of an intervertebral disc with consequent compression of nerves as they leave the spinal canal. This causes the condition colloquially called sciatica and is experienced as a sharp shooting pain radiating down the leg (when the compressed nerve is in the lower spine) or arm (when the compression of the nerve is in the neck), which is innervated by the damaged nerve. Many of these patients have had surgery to decompress the nerve, but the pain persists due to the irreversible damage done to the nerve by the compression, combined with an as yet unknown genetic predisposition to the development of chronic pain. Patients therefore complain of both leg and back pain since some of the branches of this nerve also provide sensation to the lower back, and abnormal postures they adopt due to leg pain can give rise to musculoskeletal back pain.

Other causes of neuropathic pain include trigeminal neuralgia, which affects the nerves of the face and jaw, and nerve damage due to diabetes or multiple sclerosis (autoimmune damage to the brain and spinal cord). But these groups of patients are much smaller in number. Neuropathic pain is difficult to treat using the coping strategies that are delivered in the context of a pain management program because of the episodic, unpredictable, and disruptive nature of the pain.

The second group of patients who attend the pain clinic consists of those who suffer with nociplastic pain due to episodes of inflammation from trauma or rheumatological diseases. These diseases involve dysfunction of the immune system, causing the body to try to digest itself. Repeated episodes of inflammation result in an increasing sensitivity of the pain alarm. Patients therefore experience a constant level of background pain punctuated by flare-ups when their disease is out of control. We occasionally see patients in this category with conditions such as rheumatoid arthritis or psoriatic arthropathy, but they are a very small part of our practice.

The third, and by far the largest, group of patients who are suitable for and potentially benefit from pain management rehabilitation are those who suffer with nonspecific chronic lower-back pain and patients who suffer with chronic widespread pain, usually in the form of conditions such as fibromyalgia syndrome, as we saw in Helen's case in chapter 5. These patients tend to participate in either individual sessions of pain management rehabilitation with a physiotherapist and a psychologist or in a group cognitive behavior therapy–based pain management program.

The pain management center in which I work has evolved a process for assessing patients that begins before they are seen in the clinic and attempts to outline their problem within the context of a biopsychosocial model. We receive referrals from many sources, letters written by the stationmasters at the various train stations the individuals have already visited. Most commonly patients are referred by their GP or by the GP at the request of a hospital doctor such as an orthopedic spinal surgeon or neurosurgeon. We assess these referrals and usually prioritize patients who have cancer pain or who are at risk of losing their job. Where referrals indicate that the person has not been fully evaluated and that there are outstanding medical issues that have not been resolved, particularly when they have ongoing appointments with other clinicians, we write back to the referrer to explain that it is unlikely the individual will accept the model we are proposing if they do not feel they have explored all other options.

When a decision has been made to accept the referral, we send out a packet of questionnaires for the patient to complete at home. These questionnaires include a narrative component asking the person about the onset of their pain, which often becomes very relevant if their pain started after an accident for which there is an ongoing medicolegal claim. The medicolegal process is stressful and involves multiple appointments, which can impact significantly on their response to rehabilitation. Patients often experience a great deal of anger if their persistent pain is due to an accident or event that was beyond their control—and that anger exacerbates and maintains their pain.

We ask patients about

- the sites and nature of their pain as well as the duration of their symptoms
- the clinicians they have seen and the treatments they have had
- medication they are currently receiving or have received, and whether they find painkillers of benefit or have experienced negative drug effects
- whether they are in the process of being interviewed for a disability claim, if they have filed a medicolegal claim, or if they are undergoing stress associated with filing for benefits
- whether they smoke or drink, as this has an impact on pain and is indicative of potentially unhelpful behaviors related to the management of illness
- what their expectations of the pain center are, and what their current understanding of their pain is.

That last question is often the most revealing. Responses vary from "I have no expectations" to pages and pages of ideas, concerns, and hopes related to their experience of pain. We include a diagram of the human body and ask the patient to indicate where their pain is and what sensations they are feeling. Responses here vary from a single X placed over the entire body to finely detailed and multicolored annotations indicating various pains. The diagram is useful particularly when we are looking at patients with neuropathic pain, for whom a detailed diagram indicating numbness, tingling, or burning is often a helpful indicator of a specific nerve injury.

The questionnaires attempt to capture the patient's level of distress and disability as well as their thoughts about their experience of pain.

The written element covers the biological, psychological, and social aspects of the impact of their experience of pain. The psychometric tests attempt to do the same but in a more standardized and scientific manner in order to compare patients and stratify them according to levels of distress and disability. These tests measure pain severity and interference, depression, anxiety, difficulty performing activities, catastrophizing, fear of movement, and quality of life.

The questionnaires outline the patient's degree of disability related to pain. Patients are asked whether they change position frequently to try to get comfortable and what impact pain has on their ability to walk and dress as well as turn over in bed. Patients complete a Brief Pain Inventory, which consists of a series of questions that ask them, on a scale of zero to ten, where zero is no pain and ten is the worst pain they can imagine, what their worst and least amount of pain has been in the past twenty-four hours. We also ask patients where their pain ranks on average, and how much relief medications or other treatments have provided in the past twenty-four hours. The Brief Pain Inventory includes a pain interference scale, where zero indicates no interference with daily life and ten equals complete interference; patients are asked to use this scale to rate how much pain interferes with their general activity, mood, walking ability, normal work (including both work outside the home and housework), their relations with other people, sleep, and enjoyment of life. After comparing the patient's score with the scores of individuals who have previously attended the pain clinic, we can infer to some degree the level of rehabilitation intervention they require.

Another questionnaire asks patients to rate their perceived quality of life on a scale from very poor to very good. Pain impacts function, and loss of function results in low mood; one section of this questionnaire therefore seeks to quantify the degree of depression the individual experiences, including questions such as "Over the last two weeks have you had little interest or pleasure in doing things or felt down, depressed and hopeless?" and "Have you had trouble falling or staying asleep or feel tired and have had little energy?" This section also inquires about appetite and concentration. One question specifically asks about the risk of suicide, and when patients admit to feeling that they want to harm themselves nearly every day it triggers a letter to their GP to inform them of this potential risk. The presence of moderately severe to severe depression significantly compromises the individual's ability to understand information and initiate the change needed to successfully manage their pain.

Persistent pain understandably makes people very anxious. Patients are afraid of the reason behind their continued experience of pain and can develop pain-related fear. To assess pain-related fear we administer a psychometric questionnaire that specifically explores different kinds of fears and anxieties related to pain. Patients may experience physiological disturbances due to the experience of pain: their heart may pound or race, or they may tremble when engaged in an activity that increases pain. This internal derangement and disquiet due to the experience of pain releases neurotransmitters that further sensitize the pain alarm system.

Some patients experience cognitive anxiety due to the experience of pain and are constantly thinking about pain, making them unable to concentrate. Others stop using the body part they perceive their pain to be coming from, which leads to behavioral deconditioning and physical deactivation and can spiral into disability. Avoiding movement is a natural reaction to pain and is reasonable when the pain will resolve within a few weeks, but it is unhelpful when pain persists. Muscles and joints stiffen due to inactivity, and movement then causes further pain, which reinforces fear of movement.

The most common belief about pain is that it is due to (perhaps undiagnosed) injury. Patients believe that pain is a guide, that if it hurts when they move it is because they are making the situation worse. This is not the case in chronic pain, as previously explained, but fear of movement (called kinesiophobia) severely impacts a person's ability to function. We therefore administer a specific questionnaire to establish the degree of kinesiophobia the individual experiences. The patient is given a number of statements to which they respond by either strongly disagreeing, somewhat disagreeing, somewhat agreeing, or strongly agreeing, for example, "I'm afraid that I might injure myself if I exercise," "My body is telling me that I have something dangerously wrong," "Pain always means I have injured my body," and "Being careful that I do not make any unnecessary movements is the safest thing I can do to prevent my pain from worsening."

Finally, we administer a test to quantify the patient's level of catastrophizing. Patients who catastrophize to a high degree require intensive psychological input to alter their belief system and personal ways of coping.

The completed questionnaires, made up of the patient's written answers and their responses to the various psychometric tests, give us an overall impression of the biological, psychological, and social factors impacting on the individual's distress and disability. So when Diane arrived at the clinic, for example, we had already gathered a lot of information about the way she was affected by persistent pain. Diane's psychometric tests revealed that she had high levels of kinesiophobia, informed by her own theories of why she had persistent pain: she believed that her spine was crumbling and that the MRI scan of her lumbar spine showed defects in the discs. She avoided movement in order to halt the progression of the crumbling. Due to this belief system Diane scored very high on the disability questionnaire, spending much of her time in bed and experiencing difficulties with every aspect of daily life, as indicated in her Brief Pain Inventory. Diane also catastrophized to a high degree, particularly ruminating about pain on a constant basis, and she had high levels of cognitive anxiety and pain-related fear. She scored in the moderately severe range for depression and in the severe range for anxiety. And she was a heavy smoker.

All this affected the whole family. She reported that her husband and mother had taken over most of the household chores and that she was depressed and anxious over losing her role as mother and caregiver for her children. She had a very angry son who could remember her prior to the development of chronic pain and felt the loss of his mother. The loss of her job had resulted in financial difficulties. She was looking for a quick fix for the management of her symptoms.

During the consultation Diane held her breath when attempting any movement; she would brace herself and groan when reaching for her handbag. Her husband answered a lot of questions for her—she asked him to and looked to him for answers but then got angry when he appeared to do most of the talking. She raged about the multiple injections into her spine that she had undergone without any benefit. She was angry that nobody had ever explained to her why she had developed persistent pain, and she was also angry that it had taken years for her to have an MRI scan of her spine; she believed an earlier scan would have resulted in an operation to fix her back and that her spine was now beyond repair. Diane therefore presented as a patient at the more extreme end of the spectrum of individuals we see.

The completed questionnaires are assessed by our physiotherapists, who triage the referrals into various clinics. Patients who have been sent for very specific management of medication or injections or who have significant medical comorbidities and are elderly are often triaged to a doctor-only clinic. The reason for this is that they may require some rehabilitation input, but their concurrent medical problems, such as heart or respiratory disease, potentially preclude them from the physical reconditioning that is a component of the rehabilitation within the pain clinic.

Patients who score high on the depression inventory are triaged to the doctor-only complex. They require more time during the assessment, and it is often necessary to analyze their previous interactions with other clinicians. A patient like Diane, for example, would require a thorough and detailed explanation about the nature of chronic pain. I spent a long time with her talking about acute pain versus chronic pain and essentially explaining to her how the human body works. I often talk a lot in these sessions about the pain alarm system and use multiple analogies to convey the message that the ongoing pain is due to dysfunction of that system rather than being due to ongoing damage. This message is very difficult for patients to receive and is often met with anger and disbelief. They are angry at having a condition that does not promise a cure and also at why it has taken so long for somebody to explain their condition to them. They are dismayed that they underwent treatments that, based on the theory I've proposed, clearly had no hope of helping their pain.

I explain very briefly the therapies that we offer and the strategies we employ to help people cope with chronic pain, but at that point patients are often not willing to hear what the future holds for them. The belief tends to persist that an injection, a medication, or an operation is the only therapy that has any validity—this is often the case even when patients have tried those approaches without success. When I talk about pacing day-to-day activities or mindfulness or solving problems in a more realistic way, patients become impatient.

Patients who are suitable for rehabilitation include those who present with musculoskeletal pain such as lower-back pain or other widespread chronic pain conditions such as fibromyalgia syndrome, but who have lower levels of psychological distress. They are then assessed jointly by a doctor and a physiotherapist. The power of this consultation is the interaction between the doctor, the physiotherapist, and the patient. These consultations can be fraught, particularly when the physiotherapist is introduced. Often patients who have had persistent pain for a long time have already stopped at the Physiotherapy train station and left with disappointment and unmet expectations. There is no way to exercise, massage, or manipulate your way out of chronic pain, and often patients will experience flare-ups of pain due to hands-on physiotherapy; that experience may prejudice them against the entire world of physiotherapy. They therefore view our physiotherapists with suspicion and frustration, and it is to the physiotherapist's credit that patients who are initially quite averse to physiotherapy leave feeling very reassured. The pain-management physiotherapists are experts at explaining to patients how changes in the way they perform activities can lead to an improvement in their quality of life. Patients triaged in this clinic are often not as highly distressed as Diane and are therefore more receptive to receiving the message that their condition is not curable but can be managed.

The triage process also identifies patients who can be seen by a physiotherapist working independently. These patients usually do not require any medical input, and if after the initial assessment they have any questions that the physiotherapist is unable to answer, or if the physiotherapist feels that the answers the patient requires need to come from a doctor, then they will see one of the pain consultants. While in my view the mainstay of pain management falls within the remit of psychologists and physiotherapists, the reality of the medical-industrial complex is that the word of a doctor still carries more weight with many patients.

The doctor-only complex clinics are the hardest emotionally. Usually a thorough reading of the triage questionnaires reveals how much anger, disappointment, and resistance to change the doctor may face. The written answer to the question about expectations from the pain clinic is often at the crux of the interaction, as well as ideas about chronic pain that usually relate back to what the patient has understood from previous clinicians—such as that their back pain is due to their spine crumbling. The most challenging consultations are those where the patient has been referred to the pain center by a clinician who has promised an intervention that is not appropriate.

Over the years my approach to the management of patients with chronic pain has changed. There was a time when I believed that everybody could be saved by the miracle of pain management and that if I was skillful enough in communicating information about chronic pain and how this condition could be dealt with, everybody would be able to lead amazing and productive lives. I suppose this is no different from the enthusiasm of doctors who try to communicate to their clients the benefits of healthy eating and physical exercise. I have come to accept, however, both in the arena of pain management and in medical education (as well as in every relationship I have had), that all we can do is try to be as creative as possible in communicating information and employing feedback from the people we're trying to serve.

The consultation starts with my walking out to the waiting room to receive a patient. I pay attention to who they are with and how they are walking. I can tell a lot about somebody simply by the way they walk and carry themselves. I usually invite partners into the consulting room, but sometimes the patient tells them to remain outside, and the reason for this often becomes clear over the course of the next hour. If I am with a physiotherapist, they spend some time explaining the role of physiotherapy in pain management. Then I usually summarize the results of the questionnaires the patient has completed. I have found that being able to recite details about their lives and the answers they have given gives them some comfort and makes them feel understood and listened to. And it can help build rapport. Not being listened to by their previous healthcare professionals, particularly doctors, is a major complaint of patients who attend our pain clinic.

I usually have to explain what our pain clinic offers, particularly to patients who have been to other pain clinics. I did this with Diane. I started with an explanation about chronic pain and how our belief is that the pain is due to changes within the nerves and how they function rather than due to ongoing damage. I talked about the fact that patients who have had limbs amputated continue to experience pain even though the limb is no longer there and therefore cannot possibly be damaged; the pain is still experienced because the nerve that used to go down to the limb is still present, traveling from the stump to the brain. I explained that it was very difficult for us to stop these abnormal sensations and that our pain clinic worked with people to help them live productively with the constant presence of pain rather than trying to attempt to cure the condition. I sometimes adopt this approach with patients who have a long history of persistent pain and who have had multiple negative interactions with clinicians. Clarifying the clinic's position at least gives the person the option of walking out or, hopefully, allows them to manage their expectations at the outset.

The history taken during the consultation is largely an attempt to understand the person and how they and their loved ones live with chronic pain. By the time they reach the pain clinic most serious conditions have been excluded. There are many pain consultants who argue that they are not diagnosticians and that they cannot be held responsible for missing pathology. A couple of times in the past decade I have found cancers of the spine, but in these cases the patients presented with pain and risk factors that made it patently obvious that there was something else going on. Most patients have had a thorough evaluation prior to coming to the pain center, and it is usually very clear that we are dealing with a chronic nonmalignant pain condition. This does not stop patients feeling that they need further tests; much of the to-ing and fro-ing between the rehabilitation team and the medical consultant in the pain clinic revolves around the desire of patients to confirm that nothing has been missed.

I ask patients about their pain and focus on how pain impacts on their ability to perform the basic activities of daily living, including washing and dressing as well as making breakfast, lunch, and dinner and sleeping. I ask how they spend their days. Those patients who are working (and they are often in the minority) are asked to describe how their employer interacts with them in terms of their chronic pain condition and whether adaptations and allowances have been made in the workplace. Most patients spend a lot of time watching TV, smoking cigarettes, and staring into space. This presents a challenging scenario for rehabilitation; in many ways having chronic pain has become the sufferer's occupation, and any attempt at rehabilitation threatens the homeostasis that has now evolved within the household.

Sleep is significantly disrupted by persistent pain, so we spend a lot of time focused on how the person sleeps and whether they nap during the day. Fatigue is intimately associated with chronic pain, since the brain areas that subserve these functions are both located in the same region. Chronic fatigue syndrome and chronic widespread pain often coexist, each feeding off the other. We ask patients about their mood and anxiety and specifically about acts of self-harm. We discuss medications and how useful these are in the management of persistent pain and try to identify patterns of behavior that point to addiction or dependence. We do ask about previous life events, but I try not to delve too deeply because once I start scratching the surface with chronic pain patients, issues arise that I am not qualified to deal with. A history of sexual abuse is not uncommon, as is domestic violence and difficult social circumstances.

The question that is of fundamental importance and sets the tone for further engagement is the patient's understanding of why their pain has persisted. Patients relate that surgeons have told them "I would operate, but the operation would probably kill you"; "Your discs are crumbling and there is nothing we can do because the hospital will not pay for disc replacement surgery"; "I could help you, but you're too fat." Patients may have been told that the pain is all in their head, and when I explain that chronic pain is not due to disease or injury they may think I too am dismissing it as imaginary. This belief may have been reinforced by clinicians they have seen who refer to psychosomatic pain or psychologically mediated pain. Our psychologists have had to produce a leaflet explaining the role of clinical psychology in the management of pain. Explaining pain to patients has become a significant industry.

Patients like Diane are not going to be helped by further interventions. The diagnosis I made at the end of our consultation was that she had nonspecific lower-back pain with pain-associated reduction in physical functioning, and the psychological impact in terms of anxiety and depression of having a persistent pain condition was impairing her ability to function. Biologically Diane's pain was maintained by changes in the pain alarm system, but there was no evidence on the MRI scan that she suffered with compression of the nerves in her spine. There was therefore no surgical cause or remedy for her pain.

Diane's assessment described the biopsychosocial formulation of her chronic pain condition. Targets for her rehabilitation included education on the management of pain as well as learning adaptive coping strategies from our psychologists. The aim was to reduce her fear by helping her to understand that her condition was not due to ongoing damage; this would then allow her to gradually start increasing her level of activity under the guidance of our physiotherapists. From the medical point of view, her use of opioids was destructive in that it affected her ability to think, and I suspected that at some point she had become addicted because despite their not providing any benefit she continued to use them. We would therefore have to address opioid reduction as part of her overall pain management strategy. Diane presented with clear targets for successful participation in rehabilitation, but these were ultimately our goals and our assessment, and at the end of the day it would be down to whether Diane chose to continue living in the way she currently was or engaged with our clinic in order to make changes.

There was a time when I would immediately refer somebody like Diane to one of our psychologists because of her score on the depression psychometric. A score above a certain threshold triages an immediate referral to psychology as the next point of contact within the clinic, but this number is fairly arbitrary. Some patients' low mood is perfectly understandable in the context of their pain and its impact on their function, and when I explain chronic pain to them they appear to accept the explanation and are able to see the sense in what I am offering. In these instances, I exercise my discretion and refer them to the physiotherapy team. My practice now is not to refer somebody like Diane immediately to psychology because inevitably patients who do not accept or understand the chronic pain paradigm that we operate within will not attend what is a very costly appointment.

I offer patients the opportunity to come back and see me if they have further questions about the nature of chronic pain—particularly with regard to whether they have structural problems or the need for further medical evaluation—or they can opt in to rehabilitation and see a physiotherapist or psychologist. Our physiotherapists have devised an opt-in system because of the high level of nonattenders. The hope is that there will be fewer patients who do not attend because the act of opting in implies a willingness to change.

I see new patients on Wednesday morning. In order to prepare, I spend an hour in the gym, beginning at 6:00 a.m. I also spend time in the sauna and steam room to mentally prepare for the next four hours, as there is the potential for a significant amount of drama during that time. When I lecture healthcare professionals who work in difficult clinical environments, I often use Karpman's drama triangle as a model for being aware of the potential for conflict in any given situation. Karpman used a triangle to map the various roles people play within drama-intense interactions: persecutor, rescuer, and victim. The victim feels oppressed, helpless, hopeless, powerless, ashamed, and unable to make decisions. Patients who present with persistent pain, by the very nature of the condition, often adopt this role or have developed these strategies over time. A patient who is playing the victim role may position the doctor as either the rescuer or the persecutor. The doctor-rescuer feels guilty if they don't go to the rescue, which can lead to patients having inappropriate tests or interventions; playing the role of the rescuer takes the pressure off the doctor by focusing their energies on doing something for the victim. The persecutor, on the other hand, insists that the problem is all the victim's fault and is blaming, critical, authoritative, rigid, and superior. Patients whose previous interventions have failed might switch to the persecutor role; the clinician then ends up becoming the victim. And so the drama continues.

In the clinic, I try to step outside myself to observe these interactions at a distance. I try to monitor how I'm feeling, and when I feel that I have adopted any of these roles I pause and analyze the situation. I think that most doctors struggle with not being a rescuer. Many people who go into healthcare do so because they enjoy the payoff that comes from being able to focus on helping others as opposed to becoming more integrated people themselves. I sometimes wonder if the desire to be an interventional pain doctor stems from the self-esteem boost that comes from providing therapies; even though they work only occasionally, that is often enough. I have noticed lately that this behavior manifests as doctors posting on social media platforms patient testimonials about their treatments. Unfortunately, this tends to lead to codependency; for example, giving repeated injections encourages the patient to adopt a passive approach to their pain management, aided and abetted by a well-meaning physician or one who benefits financially or psychologically from performing repeat procedures.

The corollary of Karpman's triangle is the Winner's triangle by Acey Choy, published in 1990. Choy proposed a reimagining of the three states of rescuer, victim, and persecutor. Anyone feeling like a victim should think more in terms of being vulnerable and caring and should be encouraged to accept this and become more self-aware; anybody cast as a persecutor should adopt a positive and assertive role and should be encouraged to ask the victim what it is they want; the rescuer should be encouraged to show concern and care but not overreach and problem-solve for others. As doctors, understanding which role we are being cast in within a clinical context and changing to the corresponding Winner's position makes for a more productive clinic. Burnout is an ever-present threat to doctors who work in these dramatic situations that are filled with potential conflict, and understanding our role and our own psychological vulnerability is the only way to ensure that we are able to provide good care. The alternative is the little shop in the corner of our coffee room, which is stocked by one of the nurses. I always know when one of the psychologists has had a difficult morning because they walk very fast to the corner where the chocolate is kept.

———

Diane phoned the day after we spoke. She told the receptionist, Julie, that she had read the patient information leaflet that I gave her, and while she was not 100 percent convinced by the explanation, she felt that she had run out of options with regard to her pain and was keen to improve her quality of life. Julie then triggered the referral letter to our pain management psychologist, which I composed alongside the letter to Diane's GP.

The psychologists usually see patients for around ninety minutes. Their assessments are in-depth and venture into areas of previous and current psychological functioning as well as psychosocial factors that impact on the development and maintenance of chronic pain. Individuals are usually seen alone initially; their partner is brought in toward the end of the appointment to add to the depth of the assessment. The psychologist will analyze in more detail the scores on the psychometric tests that focus on depression, anxiety, and catastrophizing along with the patient's daily routine and current sleep hygiene and the impact of pain on their social life. The psychologist will also delve into the early life of the individual and psychological injuries sustained during their formative years, which may affect their ability to cope. Previous reactions to grief and the ability to surmount these reactions are also discussed. Ultimately the aim of the psychology assessment is to identify any barriers to participation in rehabilitation, as a result of which the patient may be referred to the pain physician for reassurance or appropriate tests. The patient's suitability for group or individual work is assessed, as well as identifying their goals for rehabilitation.

After the psychology assessment Diane was referred to the physiotherapy team, who assessed her current level of functioning and whether she would fit into a pain management program. Her ability to self-care and whether she would be able to turn up daily for meetings was assessed together with her beliefs and understanding about pain and her perception and experience of the safety of movement.

The fundamental underpinning of pain management psychology is that excitatory pain pathways that send a barrage of information about harm or potential harm become active, and the inhibitory pain pathways that attempt to quell this information are compromised in the presence of anxiety, depression, threat, and catastrophic thoughts about pain. When patients learn to regain control over these emotive responses using methods such as cognitive behavior therapy and mindfulness, they are potentially able to activate inhibitory pain pathways and regain some control over the intrusiveness of pain despite not abolishing it. In the future it may be that rather than aiming to use psychological methods to teach people to accept pain and to improve their function, behavioral techniques may be specific and targeted enough to reduce the actual experience of pain. According to the authors of a paper entitled "When Pain Gets Stuck: The Evolution of Pain Chronification and Treatment Resistance," there is a need for a closer relationship between the neurobiology of pain and attempts to influence behavior. "Hurt does not mean harm" is the basis for any psychology-based pain management therapy. Overturning the idea that chronic pain is an alarm to avoid harm sits at the center of rehabilitation, and until a patient accepts that persistent pain is not a harbinger of further injury, fixed pain behaviors such as fear-induced avoidance cannot be unlocked. According to the authors of "When Pain Gets Stuck," the next generation of psychology-based medicine will move away from encouraging this acceptance of pain and will instead seek to alter that inevitability.

Current pain management rehabilitation focuses on improvement in function, psychological well-being, and quality of life despite the ongoing presence of persistent pain. We do not yet have reliable techniques to actually reduce pain through psychology-based interventions, and any attempt to link pain directly to a patient's psychology can lead to failure and reinforce the idea that the pain is all in the patient's head. The rehabilitation team employs a "five systems" model to help patients learn to manage persistent pain effectively. The five systems consider the **trigger situation** and how it interacts with the individual's **thoughts** (perceptions and predictions of the situation, silent conversations that individuals have with themselves), **emotions**, **behavior**, and **physiology** (including bodily reactions felt internally by the individual and not externally observed).

One of our psychologists who teaches this model and how it applies to pain medicine uses the example of a man who is waiting to go on his first date. The environmental trigger is the waiting. The man's thoughts might revolve around potential embarrassments, saying the wrong thing, wearing the wrong clothes. Emotionally he feels anxious. Physiologically he might experience an increase in muscle tension with an increase in heart rate, sweating, and fast and shallow breathing. His behavior during the date may be inhibited, or he might decide to abandon the endeavor altogether. If he abandons going on the date, he feels relieved, but he becomes disabled in an important aspect of life. His dating self-efficacy (defined as a personal conviction that one can successfully execute a course of action to produce a desired outcome in a situation) remains low, and he is likely to continue to be anxious in similar situations. (He is also likely to remain single.) He is therefore at risk of continued avoidant coping and ongoing impairment in his ability to date.

For somebody with chronic pain, putting on a pair of socks could be the trigger. (Other environmental triggers are fear of falling and flare-ups of pain as well as self-consciousness about pain behaviors and loss of social confidence and work.) The person may think, "If I bend over I'm going to prolapse my disc again and end up paralyzed." They experience fear and panic at the thought of performing the activity, causing an increase in muscle tension and pain; their behavioral response is to hold their breath in fear. They avoid bending over, and their partner puts their socks on for them. There is obvious relief at not having to bend and suffer the imagined risk of a disc prolapse, and the socks are on their feet (mission accomplished); however, they have become disabled in an important aspect of their self-care and tomorrow morning will feel anxious again about dressing. The action of the partner has resulted in a disabling and overprotective pattern being reinforced. The person with pain has not moved their back in a normal manner, and this contributes to further deconditioning.

Trigger situations result in feelings of anxiety and fear, leading to avoidant behavior. Avoidant behavior results in short-term relief of anxiety, but in the long term there is no opportunity to build confidence and so individuals continue to experience anxiety about situations that occur as part of everyday life. The aim of rehabilitation is to increase tolerance to trigger situations, and psychology-based pain management utilizes the concept of behavioral activation, which encourages the individual to use *approaching* rather than *avoiding* behaviors and to become active despite their negative feelings or lack of motivation. Repeated practice in the same situation results in progressively reduced anxiety with each exposure. The aim of pain management rehabilitation therefore is to help the individual to break down an anxiety-provoking activity into a series of steps. The initial steps are difficult, but once the patient builds up confidence that the situation can be tolerated, the remaining steps become easier to navigate. Situations therefore become less anxiety provoking.

Behavioral activation uses the principles of operant conditioning or associative learning, reworking a patient's association between an action and the likely consequences, encouraging depressed people to reconnect with their environment and facilitate positive reinforcement. The aim for Diane was to carry out regular and manageable as well as meaningful activities that are an important part of day-to-day function, leading to greater enjoyment and satisfaction and helping her to recover some of her previous feelings of self-worth.

An important point to note, and the most difficult one for patients to accept, is that rehabilitation and behavioral activation take place in the absence of any meaningful reduction in pain. We are conditioned to believe that rehabilitation should take place only once the pain has stopped, particularly when it comes to acute injuries. Interestingly, even with high-performance athletes and acute injury, we have moved away from waiting for a reduction in pain before rehabilitation takes place. This shift occurred when we learned that recovery can be enhanced if movement takes place early. Active recovery is now undertaken by athletes following games and might include yoga rather than rest. Medically, this technique is used in the context of patients who have had surgery for cancer, where early mobilization has been shown to reduce complications.

Diane had expressed some concerns in her initial appointment about her spine and its structural stability. Successive appointments with the physiotherapist and psychologist as well as detailed explanations about the way the spine functions had reassured her to a degree, and she was now willing to start thinking about exercise. Earlier in her life with persistent pain she had experienced good days and bad days with regard to the intensity of her pain, and this had led her to do too much on a good day; she would rush about and do all the activities she was prevented from doing on a bad day. She would then experience an increase in pain on successive days and eventually stopped having good days. She was therefore used to trying to do activities despite the presence of pain. While Diane was spending an excessive amount of time at home, she was not spending it in bed. Spending time in bed and not being able to self-care are potential barriers to engaging in a pain management program.

Diane was scheduled for our intensive program, which runs for fifteen sessions, from 9:30 a.m. to 4:30 p.m., Monday through Friday, over a three-week period. Patients are then followed up at one, three, and six months. Housing is available for patients who live far away. Patients scheduled for intensive programs are less functional and have higher levels of distress and fear-induced avoidance than is the case for those attending the standard pain management program, which runs twice a week for eight sessions over four weeks. The standard program is more suitable for local patients who have greater functionality and lower levels of distress and are more likely to be working.

At the first session of the program, I talk about the limitations of the biomedical model and how restricted we are in terms of our science and understanding of many conditions, including persistent pain. We spend some time talking about the value of X-rays and MRI scans. I describe how doctors are trained and why patients often come across doctors with different theories of why they have pain. I try to communicate that what is needed most for patients who suffer with pain, and across the general population, is a reorientation to the medical profession and the treatments we offer. I encourage patients to question their doctors when it comes to any diagnosis or treatment, and to ask about risks and benefits. Most patients, particularly in the UK, have a very passive approach to their healthcare; doctors are still perceived to be the final arbiters of decision-making in a determinedly paternalistic system.

My talk lasts about half an hour, after which the psychologist and physiotherapist take over. The aims of the psychologist are to help patients move from the medical model of pain to the biopsychosocial model of pain and disability and to educate them about the physiological effects of pain. Psychologists help patients to build up their self-management skills and to reduce their reliance on healthcare professionals. Barriers to progress are addressed and positive lifestyle changes are encouraged. Psychologists focus on the stress response and its impact on muscular activity. We use a picture of an EMG (electromyography) recording, which shows the electrical response of muscles, from a patient with chronic lower-back pain and contrast it with the electrical response of someone without pain. In individuals who do not have chronic pain there is a lack of electrical activity when the person is stationary and bending forward, whereas in patients who have persistent pain the EMG shows ongoing electrical activity throughout the cycle of movement. We use this to indicate how techniques such as relaxation training and diaphragmatic breathing can calm down these hyperirritable muscles and therefore reduce the physiological response to pain.

The psychologist will also teach the patients the cognitive behavior therapy skills I mentioned earlier, including a behavioral approach to the management of environmental triggers, as well as focusing on sleep hygiene and sexual relationships. Chronic pain inevitably spikes due to psychosocial stressors, and so adaptive ways of dealing with these flare-ups are taught. The patients' partners are invited to a session to learn what living with chronic pain is like and what constitutes helpful and unhelpful behaviors.

The physiotherapy sessions teach patients how to split activities into manageable portions and to gradually increase levels of activity. Patients are taught the importance of posture and biomechanics and are actively engaged in exercises that include stretching and strengthening as well as aerobic and pool sessions.

I will see the patients on both the intensive and standard programs again during the course, at which point we have a longer discussion about the role of medications and interventions. Much of what I tell patients is recorded in this book. When I ask them how much pain relief they have obtained from medications, they consistently answer 30 percent, which is congruent with the scientific literature. Despite this, many patients who have had negative experiences with medications are still married to the idea of taking a drug and cannot escape the mindset of seeking a pharmacological solution to their pain.

We talk about interventional pain therapies, including injections and surgery, and I explain how complex the nervous system is and how merely changing how a structure within the body looks by performing an operation—as in the case of a knee replacement or the decompression of a nerve—does not necessarily change how the nerves supplying the structure function. A patient with a beautifully decompressed spinal nerve may continue to experience their preoperative pain—sometimes more severely. Even if a knee replacement is completely technically successful and the surgery has changed the architecture of the knee, the nerves that supply the knee were already exhibiting aberrant function and therefore fail to settle regardless of how beautiful the postoperative X-ray of the knee looks to the orthopedic surgeon. That the brain constantly monitors the body for signs of deviation rather than being a passive recipient of information may explain why joint replacement sometimes does not work, as changing the biological knee for an artificial one does not change the brain's perception of what is happening to the joint. Risk factors for continuing to experience knee pain following a technically successful replacement include catastrophizing, depression, and chronic pain at another site. All knees change as we grow older, just as faces change and wrinkles appear. The changes in the knee, however, do not relate proportionately and precisely to the pain experienced. It may be that the deformation of the knee due to being overweight, for example, is what the brain responds to and what causes pain; what is needed to reduce the knee pain is in fact weight loss and strengthening the muscles that support the joint.

I find these sessions enjoyable and personally educational. I have changed my style of delivery over the years, moving from death by Power-Point to simply having a conversation with people. I spend less time talking and more time listening. It is crucial to establish at the beginning of the program the patient's ideas, concerns, and expectations of pain management from a medical point of view, or they will fail to engage.

Patients tend to be both hopeful and anxious, with a real desire for change and improvement. The medic gets only a glimpse into the transformative power of a pain management program; the psychologists and physiotherapists are the ones who accompany patients along the journey to improved function. This includes social function, so there is often an outing to the local swimming pool or bowling alley. Patients choose what activities they would like to do; for many of them this is the first time in forever that they have done anything approaching recreational activity.

I often see the patients as they check in to the program each morning. I always leave my clinic door open (when I'm not seeing patients) and watch them walk by, and over a three-week period there is a noticeable difference in the way they walk; I see a reduction in pain behaviors and a faster pace. We have not given these patients an injection or a drug—the patient simply realizes that by changing their thoughts about pain and therefore changing their behavior they can transform their ability to function.

The final interaction I have with patients in the program is an individual review, where we discuss their current medication and, hopefully, reducing it. From time to time we do have program attendees who ask us about new drugs (cannabis oil has become the bane of my life) or interventions, and it is obvious that they have not bought into the ethos of the program. The reality is that they will continue to seek a cure for a condition about which they are in denial. Then there are those who make progress but fear what will happen when they leave the comfort and support of a group. After a program ends, patients often keep in contact with one another, and a card to the team usually appears in our coffee room. It is easy for a doctor to become cynical about chronic pain management and the biobehavioral approach because we tend to see only those patients who fail to progress, but for our psychologists and physiotherapists, conducting pain management programs is a rewarding activity.

———

After Diane completed the program she reflected that in the previous year she had had a cloud filled with negative thoughts hanging over her head; she felt sleepy and lethargic all the time and had no interest in any activities. She never left the house alone. She felt overweight and miserable and was tearful for much of the time. Her activities involved internet shopping at ungodly hours, and she had difficulty walking, cooking, and climbing stairs. Three months after the program she felt happy and had lost weight (and gained it back but was attending Slimming World). She had a positive, can-do attitude and had regained her interest in life. She was able to leave the house and enjoy social occasions and activity. Six months later she wrote that she was happy with life and had lost weight again. She now feels able to make decisions about the future, is considering working, and is even more physically active.

Diane's story shows the positive effects of pain management programs, but these programs are not a panacea. Chronic pain is exactly that—unrelenting. Human beings struggle with diseases that require active self-management and a lifelong commitment to wellness. The difference in outcome between two patients is often not medical input but the person's commitment to managing their condition. One UK interdisciplinary pain management center that tracked patients in a residential pain management program found that clinically significant gains on pain intensity were achieved by 19 percent of patients immediately after the program and 17 percent at a nine-month follow-up assessment. Improvements in psychological measures such as depression, catastrophizing, and the confidence to manage pain (self-efficacy) were achieved by 55 percent of the patients at posttreatment and 44 percent at a nine-month follow-up assessment. This means that no improvement in pain intensity may occur in 71 percent of patients immediately after treatment and 83 percent at follow-up. No improvement in psychological measures may occur in 45 percent of participants at the end of the program and 56 percent at nine-month follow-up.

There are no studies looking at the long-term outcomes of patients who have participated in a pain management program. The detractors of this form of therapy (usually doctors involved in interventional pain therapies) cite the lack of long-term follow-up as a reason for persisting in trying to achieve pain *relief*. As with all things in life, the truth lies somewhere in between. Unfortunately, both camps tend to be polarized in their points of view and stick religiously to the fundamental tenets of their paradigms.

I feel more at peace these days when I sit in my new patient pain clinic. I am less the George Michael of "Club Tropicana," "Young Guns," and "Careless Whisper" than of "Faith," "Older," and "Heal the Pain." This peace and acceptance of my role has come after a long time. When I started as a consultant, I believed that I could save everybody by selling them interdisciplinary rehabilitation. I felt that I could bring everybody back to psychological health and back to the light. I thought that I could get everybody with chronic pain to successfully pace their activities, to engage with and understand pain, and to overcome it with newly discovered resilience. I was a true and indiscriminate believer. I would often get quite cross with one of our physiotherapists who has been around for a while because she didn't seem very hopeful about most patients. In retrospect I realize that her experience has given her a very clear sense of those patients who will engage with pain management and those who will not. The reality of a pain clinic is that, according to our outcome measures, we actually help about 10 percent of people who suffer with chronic pain to move forward with their lives, while the rest find some way to live with their condition, usually in a way that we would consider not adaptive or contributing to society but that works for them.

Chronic pain cannot as yet be cured. Neither can diabetes mellitus or high blood pressure, for that matter. Our aim therefore is to focus on reducing the distress and disability of a life of persistent pain. Multimodal pain management—collaborative and integrated use of needles, narcotics, and knives along with words and understanding—probably represents the kindest, most effective, and least harmful approach to the management of the disease that is chronic pain. Interdisciplinary collaboration—psychologists working with doctors working with physiotherapists working with occupational therapists and nurses—where the biological and the behavioral are addressed by a team operating with the same philosophy, seems to be the best way to deliver this care.

CHAPTER EIGHT

Torpedo Fish

I feel as if someone is putting out a thousand cigarettes on my leg. Sometimes it feels as if Skittles are being thrown onto my leg, and no matter which part of the outer side of my leg I touch the sensation is always felt somewhere else.

I don't know if you have children, Doctor, but imagine that you wake up in the middle of the night to go to the toilet and you have to walk down a dark passage in your home to get there. You had a long week and probably too much to drink the night before. You didn't get the promotion that you went for and so probably won't be able to afford the holiday you promised the children. One of your children has left their Lego in the middle of the corridor and you step on it. The awkwardness of the shape of the Lego as it bruises your foot and the horrible numbing, electric-like *pain is how my leg feels all the time. I feel as if hundreds of Lego pieces are being pressed into my leg all the time and the disappointment of the children and the lack of promotion and the feeling of weariness never stops.*

The burning never stops. Sometimes during the day I can forget it for a little while, but at night when everything is quiet it returns. I toss and turn, and I cannot bear to have the bedclothes touch my legs. I know it's my fault, I know I should have managed my diabetes better.

These testimonies are from patients who have neuropathic pain, which I have touched upon in some of the preceding chapters. Neuropathic pain is pain due to disease or injury of the system that registers and reports sensation (nerves) and affects 6 to 8 percent of the adult population. When you have nonspecific lower-back pain—nonspecific in the sense that the pain is not linked to a fracture of the spine or compression of a nerve or inflammation of the joints—the pain tends to trouble you in a predictable pattern, perhaps when you sit or stand for long periods or lift something heavy. Patients can usually therefore adopt strategies to minimize pain. They generally tend to be okay when pacing gentle activities such as watching television; pain becomes a problem only when they exceed their usual activity levels. For patients with nerve damage the experience of pain is materially different: the pain is unpredictable and consists of odd sensations, which doctors refer to as positive and negative symptoms. A positive symptom would be something like an abnormal feeling or sensation, called a paresthesia, which when experienced as unpleasant is called dysesthesia. Patients also experience tingling and burning, as described in their testimonies. Negative symptoms are when patients experience the disturbing paradox of feeling pain even though the area affected is numb.

Nerve pain tends to affect a specific body area because nerves give sensation and power to certain areas. Where the peripheral nervous system is affected (from traumatic nerve damage or sciatica or diseases such as diabetes), this usually involves the legs or arms. The peripheral nervous system consists of all the nerves that radiate from the spinal cord and from the brain but does not include the brain or spinal cord themselves, which are referred to as the central nervous system. Central nervous system disorders such as multiple sclerosis and strokes, it could be argued, cause even worse symptoms, and patients may have severe debilitating neuropathic pain which is also accompanied by deficits in the function of their muscles (with associated spasticity and musculo-skeletal pain), and often their bladder and bowel don't work properly.

Nerve pain can mimic other non-neuropathic pain conditions; we therefore take great care when diagnosing this problem. The patient must have a disease or injury of a nerve that can logically explain their symptoms. They must have both positive and negative symptoms and signs such as decreased sensation to cotton wool or a pinprick or increased pain and sensitivity within the distribution of that nerve terri-tory, which doctors look for on examination. If we are still not sure, nerve conduction studies can be ordered to confirm the diagnosis. These tests consist of placing needles into muscles and measuring the electrical activity along the nerve to assess if the nerve has been injured. It is possi-ble, however, to have nerve damage but still register normal electrical results, as the tests are simply not sensitive enough to detect the damage that has occurred in some cases.

The most common neuropathic pain conditions that I see in the clinic are from prolapsed discs that have damaged the nerves as they leave the spinal canal either through a chemical irritation or direct com-pression. We still don't know what the precise evolution of long-term pain from a prolapsed disc is, and there are patients with disc prolapses identified on MRI scans who never develop persistent nerve pain. We also see patients who have nerve pain from diabetes, where chronic high blood sugars have destroyed the blood vessels supplying oxygen to nerves, which then die and regenerate haphazardly, resulting in pain in the arms and legs—often referred to as a glove-and-stocking distribution of pain. Most neuropathic patients are resigned to a life on anti-depressant medication and antiepileptic medication to treat their nerve pain, and I have already described how poor these medications are at relieving pain. At best patients may get 30 to 50 percent pain relief, often at the cost of an impaired ability to concentrate and function; they are transformed into the walking dead, shadows of their former selves. Inevitably, due to their complaints of persistent pain and distress, their GP starts them on opioids, which escalate over time. This opioid haze does not improve their experience of pain but will further affect their ability to function.

Of all the patients I see in the pain clinic, those who suffer with neuropathic pain are probably the ones that have the best chance of finding a biomedical solution for alleviating their pain. Most of the patients we see in pain clinics are treated with a combination of medication, pain management interventions (steroid and local anesthetic injections or the burning of nerves), and pain management rehabilitation, discussed in the previous chapter. Peripheral neuropathic pain, however, is one of the few conditions for which we can successfully modulate the abnormal information emanating from damaged nerves through electrical stimulation of areas of the spinal cord. When I talk to patients about this treatment, they look at me as if I am the reincarnation of Mary Shelley and the therapy I am offering straight from *Frankenstein*. The reason for this is because it involves modulating the nervous system with the application of an electrical current.

───────────

The first recorded effect of electricity on the human body dates back to 2500 BCE. A picture illustrating a fisherman catching a Nile catfish and experiencing what looks like a painful shock was found painted on the wall of an Egyptian architect's tomb. Some two thousand years later, Aristotle mentions the effect on other animals of torpedo fish, also known as electric rays, and it was recognized that the shock from the torpedo fish could be transmitted through a trident used to spear the animal. The numbing effect of the torpedo fish was used satirically to describe the effect of the conversation of Socrates on one of his pupils. It was a Roman physician called Scribonius Largus in the first century CE who first described the medical use of the torpedo fish when he cured Anteros, a court official of Emperor Tiberius, of his gout. He suggested that the patient should stand on the shore washed by the sea until his whole foot and leg up to the knee was numb from being stung by the resident torpedo fish. Avicenna's text *The Canon of Medicine* recommends the use of torpedo fish to treat headaches and melancholy and for arresting epileptic seizures. The phenomenon of electricity was obviously not yet understood, however; one Roman physician suggested that the torpedo fish was releasing a venom.

Nearly two thousand years later, in the eighteenth century, Henry Cavendish constructed an artificial torpedo fish using devices for storing electric charge in a leather case resembling the size and shape of the fish. The flurry of experimentation into the phenomenon and subsequent increased understanding of the nature of electricity in this century resulted in an explosion of electricity-based devices used as medical therapies. Electrotherapy was put forward as a panacea for use by the medical profession but was also utilized by charlatans to exploit the vulnerable. It is probably this legacy that most influences patients' skepticism toward my proposal to use electricity to manage their pain.

The first electrical device designed specifically to manage pain was patented by Charles Willie Kent in 1919 and manufactured in Illinois; it was called the Electreat and was probably similar to a transcutaneous electrical nerve stimulator (TENS), which was developed much later and marketed as a treatment for every imaginable illness. The gate control theory by Melzack and Wall (discussed in chapter 2) was fundamental to the development of electrical pain management, as it proposed that pain perception is not (as Descartes had described) a line-labeled pathway that continues uninterrupted from the source of the injury to the brain; instead the spinal cord acts as a gating mechanism that can be influenced by pain perception. This paradigm shift in understanding pain opened the way for the development of electrical therapies to manage pain. Melzack and Wall proposed that the gate could be opened or closed depending on the balance between the stimulation of small and large nerves. When you step on a Lego piece, multiple behaviors that can influence the gate are activated: taking pressure off the foot, sitting down and swearing, and rubbing the foot all play a role in altering pain perception. Rubbing stimulates large fibers in the foot, which compete with the smaller fibers transmitting the electrical stimulus encoding harm. The birth of the TENS machine was a direct result of this theory: by applying electrical stimulation to electrodes taped to the skin of the painful area large fibers are stimulated. When I explain spinal cord stimulation (SCS) to patients, they often ask if it is similar to TENS.

Implantable devices to stimulate nerves were initially placed near the peripheral nerves found superficially in the arms and legs. In 1967 Norman Shealy from the Western Reserve School of Medicine had the idea to stimulate the large nerve fibers as they gathered in the posterior columns of the spinal cord. Instead of stimulating these fibers through the skin peripherally, he believed they could be accessed directly where they bundled in the spinal cord. Theoretically this would enable a larger area of the body to be stimulated (more nerves servicing a larger area are found tightly packed in a small part of the spinal cord) than could be achieved by merely placing electrodes on the skin. The first model of the spinal cord stimulator involved an internally placed lead that entered the cerebrospinal fluid and required an external power supply connected by a needle through the skin. Modern spinal cord stimulators are placed in the epidural space to avoid the complications that occur when the lead floats in the cerebrospinal fluid, and they do not require an external power supply. The first patient to have a stimulator implanted had cancer; the SCS was used to successfully treat this person's cancer pain for four years.

The use of electricity to manage pain developed alongside the use of electricity to manage cardiovascular disease. The first spinal cord stimulators were essentially modified versions of the electrical devices used to treat hypertension. As devices developed initially, they consisted of an electrode with its antenna. The patient had a battery-operated power supply and controls that communicated with this antenna. In 1981 Medtronic, the company Shealy founded, produced a fully implantable device.

In the 1980s electrodes with multiple contacts were introduced, which meant that different parts of the spinal cord could be stimulated, covering a wider pain area. In 2005 the first rechargeable battery for use in SCS was invented; batteries would need to be replaced less often and more energy-demanding therapies could be used. Between 2005 and 2007 several landmark studies in SCS were carried out, demonstrating that, for neuropathic pain due to lumbar spine problems, the therapy was more cost-effective and produced better pain relief than further surgery. The therapy has been approved by NICE, which issued formal guidance for its delivery in 2008 and reviewed it again in 2013, and it is supported globally by the International Neuromodulation Society.

Many studies have looked at whether the initial costs of the spinal cord stimulator are justified. At present, these devices cost north of £15,000 in the UK (prices vary widely in the US), but it has been shown that this initial cost is worth it given the potential for avoiding repeat surgery or ongoing investigations, injections, and drug therapy.

Patients' apprehension when I broach the subject of SCS is palpable. I often marvel at how patients will happily consent to the most horrific and invasive surgery and will agree to take mind-altering substances, but the idea of inserting a device that is able to generate a current in their body provokes immense distress (even though we are made of water and electricity). Electricity frightens people despite its silent presence in ensuring that our societies function—Frankenstein, homemade medical devices, and the electric chair play a role in the unacceptability of electrical therapy. It is interesting, though, that patients will readily accept the use of a pacemaker. I suppose the reason for this difference is that without a pacemaker or an implantable cardiac defibrillator (which can deliver a shock should your heart stop), you will probably die. In the management of pain there is always the choice to live with your pain or not. As I remind trainees, chronic pain doesn't kill anybody, and therefore the acceptability of a therapy is always based on a careful assessment of risk and benefit, on the part of both the patient and the clinician. I remind myself constantly that any patient who comes into the clinic, no matter how distressed, was living with their chronic pain the day before their appointment and will continue to live with chronic pain the day after, which makes me less likely to adopt the role of a rescuer in response to their playing the role of victim.

Implanting a permanent device into a patient demands that the assessment process and the long-term care of the patient are handled by a neuromodulation team, formed from a subset of members from the wider pain practice. The team must be able to assess the complexity of the patient and the therapy as well as the factors that will impact on the success of the treatment. First, patients are seen by me or one of my colleagues who implant these devices. Second, they are evaluated by our specialist spinal cord stimulator nurses in an education session to introduce them to the therapy and the technology. This assessment is designed to decide whether the patient will be able to cope with the programming device and with the trial period. Spinal cord stimulation is a relatively straightforward therapy, but a degree of dexterity and technological savvy is required to manage the technology, including the ability to understand and adjust the programmer. The patient's social situation and their ability to care for their surgical wounds is given consideration, and our nurses begin a conversation about rehabilitation and improving function.

Third, patients are assessed by a clinical psychologist. The aim of this appointment is to position the patient's engagement with SCS within the wider context of improving their function and reducing the distress and disability associated with persistent pain. We generally find that patients who are poorly oriented toward improving their quality of life and function do less well with the therapy. We present SCS as a quality-of-life-improving therapy rather than simply a cure for pain; while the aim of the therapy is primarily to reduce pain, if the patient does not understand that rehabilitation is the ultimate goal (because we cannot completely obliterate the abnormal sensations), then inserting the device does not achieve the desired outcome. Patients who have high levels of catastrophizing and uncontrolled psychological problems such as psychosis or depression also respond poorly; in my view this is simply because their brain is overwhelmed and cannot appreciate the change in sensory information reaching it as a result of the spinal cord stimulator's effect.

The psychologist also assesses how patients will cope if the device fails to relieve their pain. Patients often come to us with great expectations (usually because of promises made by a correspondingly distressed surgeon when referring the patient to the pain clinic), and we try to manage and anticipate potential distress should the therapy fail. We explain that their ability to manage the situation psychologically can also influence the sensations they experience. Our psychologists often unearth issues we clinicians are not privy to; for example, if a patient discloses a substance misuse disorder or confides that they have contemplated suicide, the psychologist will sometimes recommend rehabilitation therapy prior to a trial of SCS to give the patient the best chance of having a good outcome.

Once all three teams have assessed the patient, a decision is made as to whether to proceed with a trial of SCS.

Most patients are referred by our neurosurgical or spinal orthopedic colleagues who have operated on the patient's spine and, from a technical point of view and based on a repeat MRI scan, have done so successfully, yet the patient continues to experience pain. Other patients are referred by their GP or by a pain consultant from a practice that does not insert the SCS devices, possibly because they do not understand them, although we have made efforts to try to communicate our outcomes to our colleagues. I sometimes think that the clinicians and clinics that refer patients for this treatment are reluctant to do so because it means admitting that their practice does not embrace the breadth of pain medicine. In any event, our struggle to raise the profile of this therapy, which probably has the best evidence base in all of pain medicine, continues.

Our role as doctors at the point of first contact is to assess whether the individual has the type of pain that will be responsive to SCS: neuropathic pain. This is not always straightforward; just because you have had a discectomy does not mean that the pain you had when you first talked to the surgeon was due to damage of a nerve. (A disectomy is the partial removal of an intervertebral disc, which acts as a shock absorber between the individual bones of the spine but which can prolapse and compress a nerve as it leaves the spinal canal.) Nonspecific lower-back pain with trigger points that form in muscles can mimic the pain from sciatica. Patients consequently go on to have surgery on their spine when they never actually had neuropathic pain. We also assess whether the patient is medically suitable for a trial of SCS. Uncontrolled diabetes, heavy smoking, and reliance on high-dose opioids all pose risks of possible infection and morbidity from the procedure. We assess whether patients are able to lie flat for the forty-five minutes that it takes to insert the device; depending on the degree of disability that patients develop due to their pain, this may not be possible. And as mentioned earlier, we check that the patient is oriented toward improving their quality of life.

Doctors have a duty to inform the patient of their personal experience with a treatment and what their personal complication rates are, so at this initial meeting I always explain my entry into, experience with, and practice in the context of the history of the development of SCS. As doctors we are involved with people at a level where trust is vital and transparency in the face of their adversity is paramount to ensuring satisfactory outcomes. Competence, caring, and kindness should pervade medical care (and life in general).

I was first introduced to SCS in 2007 when I was a trainee at the hospital where I currently work. I often reflect these days—twenty-two years since graduating from medical school—on how most of my life-changing decisions have come about because of love. I came to the United Kingdom to keep a romantic relationship alive, and I ended up becoming a pain consultant as a result of the pain from the loss of that relationship. It occurred at a time when I was doing my intermediate training in pain medicine and I was at a particularly reflective time in my life, looking to engage long-term in understanding myself. In 2007 I was in a relationship that required me to stay in Manchester, and I was told that if I wanted a job at the hospital I would have to learn how to insert and manage SCS devices. So I did.

When I started implanting spinal cord stimulators in 2007, we used systems that relied on producing a pleasant tingling sensation (paresthesia) in the area where the patient usually experienced pain; this is known as paresthesia-based therapy. The original idea underpinning SCS was that if you were able to produce a tingling sensation in the affected area by placing the electrode over the area of the spinal cord where the nerves that represent that area were situated, that would result in pain relief. The amount of pain relief was directly proportional to the degree of coverage of the painful area. These therapies used low-frequency currents at low amplitudes. You could adjust the pulse width and the amplitude in order to increase the electrical current's depth of penetration into the spinal cord or broaden how many nerve fibers it included. Imagine a wire four inches long with a bubble around it; changing the shape of the bubble will increase the amount of coverage of the patient's painful area. Too big a bubble, however, will catch the nerves that exit the spinal cord carrying both sensation and movement, which will result in painful muscle contraction. Another problem we had with these paresthesia-based therapies was that when the patient moved, the spinal cord would float toward or away from the implanted electrode, changing the area, degree, and intensity of stimulation the patient felt because of changes in electrical resistance. The companies involved in manufacturing SCS systems tried to overcome this limitation by producing batteries that could sense the person's position in time and space. When they were upright you could program electrical variables that adjusted for the resistance in that position, and you could do the same when they were sitting or standing. This was an attempt to maintain a constant degree of stimulation despite body movement. It worked a bit like a phone whose screen changes from landscape to portrait orientation depending on the way you hold it. The other challenge, alongside producing constant paresthesia, was to make sure that the device was compatible with MRI scans, which the patient may need in the future given the increasing usefulness of this technology in detecting, for example, brain hemorrhages, ligament damage, and breast cancer.

With any new patient the therapy is first trialed before a permanent implant is put in. For reasons that we still don't fully understand, not everybody responds to this therapy. There is some suggestion that the longer a person has had neuropathic pain, the less likely it is that we can influence the damaged nerves. I tell patients that their pain is a bit like a bowl of soup: when the soup is fresh and hot, adding salt and pepper can definitively change the flavor; when the soup has become cold and congealed, though, it is less likely that the addition of salt and pepper will influence the flavor. Unfortunately, we have no way of reheating the soup. The great sadness is that despite the vast evidence for this therapy and its longevity, patients often come to us after having had their pain for more than four or five years. This holds true across the UK, and it represents a significant failure to integrate the therapy more effectively into the management of a condition that causes a greater impact on health-related quality of life than cancer, heart disease, or respiratory disease.

I mentioned before that the aim of SCS is to produce tingling in the affected area in order to provide pain relief. The trial procedure therefore involves placing the lead and subtly changing its position up or down and from side to side over the spinal cord in order to target 100 percent of the patient's painful area. This can take a long time in the operating room, and once the patient stands up, the lead can move in relation to the spinal cord, or the spinal cord itself, being a mobile structure, can alter its position. The painstaking exercise of mapping stimulation to the patient's painful area can therefore be lost in the recovery room.

Spinal cord stimulation has evolved as a therapy, and much labor has gone into removing such problems in order to achieve paresthesia that is predictable and consistent. Better anchors for the leads prevent them from migrating up or down. Batteries are smaller and rechargeable and MRI-compatible. Programming uses algorithms able to more quickly target the nerves responsible for sensation to a particular area of the body and can focus the electricity on a particular area, such as the foot or the knee. Still, despite these attempts, over time a percentage of patients started to dislike the tingling and started obsessing about the tingling rather than about their pain. The change in tingling with movement was inhibitory to increasing activity (which is, after all, a large part of the desired outcome). Patients were advised not to drive when their tingling was activated for fear that a sudden uncomfortable burst of stimulation could distract them and cause an accident. Nor could patients sleep with the therapy on, which was a major impediment because pain is often worse at night.

Around 2008 rumors began to surface in the SCS community of a new therapy that did not rely on producing tingling to achieve pain relief. Studies were produced that followed up patients who had been implanted with these devices, and the results were hard to believe because they were so good. These were prospective studies in a series of patients, and they were done in centers across Europe by clinicians who are well known for being early adopters of therapies. The idea of SCS using tingling was rooted in the very believable concept of the gate control theory, and the idea that you could have pain relief through a lead implanted in the spinal epidural space that did not rely on tingling was heresy—particularly since the company making the new devices could not provide a clear mechanism of action or explanation for why they should work.

Much work has been done in traditional SCS in order to elucidate the mechanism of action. This was studied by injuring the sciatic nerves of rats, then implanting spinal cord stimulators to assess what was happening at the level of the spinal cord. These experiments found that wide-dynamic-range neurons, which are made more excitable in neuropathic pain, are affected by the activation of the touch fibers. There is also evidence that the stimulation of the spinal cord results in the release of serotonin and noradrenaline, which, as we have seen, inhibit the excitable neurons that give rise to pain. Being told that you could achieve pain relief without producing tingling was very much like being a sixteenth-century Catholic priest told that the Earth revolves around the Sun rather than being the center of the universe. It was much easier to believe that this was a placebo effect rather than due to a definitive therapeutic benefit.

I have to say that I was on the side of the skeptics when it came to adopting what is now known as high-frequency SCS. This new system, made by a company based in San Jose, California, uses 10,000 hertz as a frequency. The idea is that this high-frequency current changes the balance of misbehaving nerves in the spinal cord. If we think of the nerves as a class of children, the current empowers those children that behave well by increasing their power, which in turn exerts an inhibitory effect on the badly behaved children and therefore makes the class overall seem better behaved. The tingling-based therapies, on the other hand, create a distraction outside of the classroom, taking attention away from the badly behaving children.

I developed a relationship with one of the representatives from the company that produced the 10 kHz paradigm. Over the years he would bring me the results of prospective studies and I would continue to doubt. Unless he showed me a mechanism of action or a study that demonstrated the superiority of the new therapy over the traditional system, I could not in good conscience offer it to my patients.

In 2015 the results of a landmark randomized controlled trial were published, and a follow-up study of the same cohort of patients was published two years later. This study represented the largest randomized controlled trial in the history of SCS. It demonstrated that the 10 kHz device was in fact superior to paresthesia-based therapies in relieving back and leg pain in patients with peripheral neuropathic pain due to spine disease.

There was much unhappiness and gnashing of teeth at these results, particularly from the companies that had been manufacturing the traditional spinal cord stimulators, and a rapid scrabbling around as they tried to come up with an alternative narrative to what was fundamentally a complete market disruptor. In much the same way that the Catholic Church went after Galileo when he challenged the prevailing model that put the Earth at the center of the solar system, the companies producing traditional devices went on full attack. At the center where I work, my colleagues and I sat down quietly and reviewed the scientific paper. We considered how the study was conducted and concluded that based on this evidence we had no choice but to offer patients the therapy. We implanted our first patient with a 10 kHz system in 2016. We have to date implanted over a hundred patients and our results have been consistent with the published evidence on pain relief. And we no longer spend upward of an hour and a half in the operating room mapping tingling; the procedure takes about thirty minutes (meaning we are able to implant more patients). Because patients do not have to program the device according to their position, they are able to have it on when they are sleeping or driving, and because it is not affected by the patient's movements it offers no obstruction to their rehabilitation goals.

Paresthesia-free or paresthesia-independent SCS, which is what the 10 kHz system represents, has its competitors. Next to come along was burst stimulation, which purports to have a different mechanism of action whereby it reduces the level at which an individual feels stimulation to just below perception. Newer systems, where the lead is placed around the dorsal root ganglion (the exit site of the nerve as it leaves the spinal cord) rather than the spinal cord, have been produced in order to focus tingling more effectively. Companies have come up with their own version of paresthesia-free programming, referring to the total charge delivered rather than the frequency.

The 10 kHz spinal cord stimulator is a great example of a market disruptor. Its introduction is no different from the introduction of fruit-named telephones versus those that were produced in Finland. If you produce a therapy overnight that sweeps away the problems associated with the previous therapy, you can end up with a whole industry in turmoil. People's livelihoods depend on these therapies, and their behavior is dictated accordingly. Relationships are established between companies and clinicians who offer a therapy, and these relationships sometimes persist in the absence of evidence to support their ongoing use and in the presence of evidence to support the use of a different therapy. NICE has been caught in the middle of this battle, as have various societies that represent pain clinicians, with everyone producing furious arguments and counterarguments. Allegations in whispered tones have even been made about spinal cord injury due to the 10 kHz frequency charge in order to detract from the delivery of this paradigm-disrupting therapy. In our unit, we quietly continue to collect data following the implantation of 10 kHz devices, not only on pain relief but also on improved sleep, function, and reduction in medication use. We inform our patients of potential outcomes based on these data compared to the results of the published studies. In our view, all we can do is be transparent about our data and continue to monitor the efficacy of the therapy beyond just pain relief.

The ability to engage more actively in rehabilitation without the distraction of paresthesia, as well as being able to sleep and drive, are of priceless importance to patients. The 10 kHz therapy does not require extensive programming, so we have less cause to bring patients back to the clinic, which is more convenient for them and for us. Spinal cord stimulation does transform people's lives. It is probably the most satisfying therapy a pain clinician can deliver, and patients say it has saved their lives. It is not an instantaneous therapy; there is an initial period from when the nerve is being modulated to the point where the patient starts noticing pain relief, which makes us believe that it is not a placebo effect. In addition, the 10 kHz platform can produce the traditional paresthesia-based stimulation and a form of burst stimulation. One system to rule them all.

I insert spinal cord stimulator trial leads on Monday morning. It is the best part of my week. I know that there is at least an 85 percent chance that I will provide the person lying on the operating table with more than 50 percent pain relief, on average. Some patients get close to achieving 100 percent pain relief.

The implantation takes place at an outpatient unit. The patient is awake during the procedure, which is done under local anesthetic. Many patients are anxious about being awake and concerned about whether they will be able to lie flat. For a long time we used to give patients intravenous sedation and pain relief in order to overcome the inevitable anxiety of having the procedure; now I find that a small oral sedative prior to coming to the operating room helps patients with anxiety.

We use an X-ray machine to place the needle in the epidural space. We thread a wire through the epidural needle, and, for patients with leg pain, the final lead placement is around the area where a woman's bra strap would be. The exit site where the wire comes out of the skin on the side of the torso is just above the patient's beltline, and the wire is then attached to an external battery. I usually tell patients all sorts of stories while they are being operated on, and we discuss politics or the week's events or soccer or what the patient has done that weekend; this is designed to distract patients and to enable them to focus on the future. Our SCS nurse sits at the end of the bed and also talks to the patient, particularly at times when the procedure may be more uncomfortable. Behind me is the representative from the company that manufactures the device; he is there to check that the electrical connections have been made appropriately.

Our SCS team also includes a neurosurgeon. He provides us nonsurgeons with a surgical governance framework. One of the greatest difficulties in the world of neuromodulation is that traditionally implanters have been drawn from a subset of needle jockeys within the world of pain medicine who are trained as anesthesiologists. We anesthesiologists are primarily trained in facilitating surgery by anesthetizing patients and providing them with pain relief. We usually take a background role, so being at the center of a surgical procedure is unusual. It can confuse the staff when sometimes I'm at the anesthetic machine minding my own business and at other times I am the one doing the operation. We anesthesiologists are generally quite risk-averse and like to live in a world where we control all the elements; for us, failure results in an almost immediate and catastrophic outcome for the patient. Surgeons have a different temperament; they understand the "slings and arrows of outrageous fortune" of life as a surgeon and can, over time, develop the ability to deal emotionally with complications.

Some units therefore operate a trial system whereby the pain doctor inserts a temporary lead, and once the trial is completed the lead is removed. The patient then has to see a neurosurgeon, who performs an open and invasive operation under general anesthetic in order to place a new lead and a permanent battery. Sometimes the patient complains that the outcome following the removal of the trial lead is not the same with the new surgical lead. In that case the anesthetic-trained pain doctor is skilled at inserting the percutaneous lead, but the surgeon is not. There are some units whose surgeons have learned how to put in percutaneous leads and therefore do their own trials and insert their own batteries. There are also units, such as ours, that are run entirely by pain doctors who are trained anesthesiologists, who insert a permanent lead at trial stage, which then remains in place and is connected to the internal battery that we insert after the trial is complete. There is no standardization in the delivery of neuromodulation.

There has been a great deal of progress in neuromodulation in some countries over the past few years. Some have implemented attempts to standardize education within the field and established minimum acceptable standards of technical competency among implanters. They have also tried to actively promote women in neuromodulation, as the number of women working in this area, and in pain medicine generally, does not reflect the more than 50 percent enrollment of women in medical school and the high number of women in anesthesiology, where pain medicine draws its trainees from. The uncertainty about the future of pain medicine as a specialty due to the decommissioning of interventional therapies will not, I fear, help this situation. Many will opt to just push propofol (an anesthetic drug) for a living.

Following surgery, patients are sent back to the outpatient unit, where the device is programmed, and are given instructions for the next ten days. They keep a diary of their pain scores and how they feel during the trial. I see them again on the following Wednesday, but our nurses first phone them on Friday, by which time we usually know whether or not the trial has been successful. Spinal cord stimulation with paresthesia-based therapies is a bit like Marmite—patients either love it or hate it—but with the 10 kHz device there is resigned disappointment if the therapy fails to deliver at least 50 percent pain relief. Those patients who experience pain relief find the therapy transformative. Their partners often say they have been less grumpy, and the patients sometimes say that for the first time they have been able to feel their feet again and are able to walk without a cane. It really is the most gratifying therapy to be involved with.

If the patient has achieved more than 50 percent pain relief, then ten days after the insertion of the trial we cut the external wire. Some patients find this quite distressing because they know that in around forty-eight hours' time their pain will have returned. We then insert the permanent battery the following Monday. The surgical procedure is the same, only this time a small pocket is made in the patient's back just above the buttocks for the battery. The battery measures approximately 1.5 x 1 inch and is about 0.4 inch thick. It is inserted under the skin and, depending on how fat the patient is, may or may not be visible. Patients take about two weeks to heal. After that they have a permanent device (and a special card to present at airport security).

We see the patients two weeks later to check that their wounds are clean, and then we see them at three, six, and twelve months. We therefore witness a gradual transformation in somebody who was not able to do very much, was depressed and potentially going to lose their job. One of my patients used to be a rugby coach; when I first saw him he had lost that position and the social life that came with it. He felt unable to participate in day-to-day activities and was struggling at his workplace. Initially he was quite unsure about whether a spinal cord stimulator was for him. He went to see a pain management physio-therapist who reassured him that his spine was structurally safe and got him to start exercising, but he continued to struggle with the neuropathic pain in his leg. Eventually he decided to have the SCS trial. He had a very successful trial, and when I saw him the other day in the waiting room, he hurried up to me and said that he was coaching rugby again. He has regained his sense of humor and is back to doing what he loves. A little bit of electricity has facilitated this change.

Another of my patients is a policeman. He sustained an injury at work and was unable to walk because of constant pain. His mobility was improved by surgery, but his leg pain continued. He struggled for years taking medication that did not help with his pain and affected his ability to think and concentrate. He struggled to drive, and he gained weight because of his inability to do any exercise. Spinal injections every six months took a fraction of the pain away, but they were never completely effective. After a trial of SCS he has now lost over forty pounds. Apart from the very small scars on his back you wouldn't even know that he had a device. He has gotten over 90 percent pain relief and hasn't had to take any time off work since having the device put in.

<hr>

In 2015 I was invited to the headquarters of the company whose paresthesia-based systems we had been using for many years. The aim of the visit was to meet with their scientists to discuss the therapy and to tour the factory where the equipment was manufactured. I had grown up with a penchant for watching TV programs about how things were made and found the factory tour illuminating. It gave me a better understand-ing of the challenges in manufacturing the equipment and the rigorous safety standards involved. I also met the inventors of what I think is the world's greatest spinal cord stimulator lead anchoring system and visited the 3D-printing lab where they devised the equipment. Despite no longer using the rest of the company's equipment, I continue to use the anchor.

As part of the visit to the factory I attended the annual meeting of the North American Neuromodulation Society (NANS). This meeting is always held in Las Vegas because, I was told, even the smallest city in the US has a direct flight to Sin City. I had never been to an American medical meeting. They are so much more fun than any European or British meeting and massive in scale. I sat in the audience of the opening ceremony—and I call it an opening ceremony deliberately because it was like the Olympics—when the lights suddenly dimmed and triumphant rock music blared. The organizers of the meeting stepped onto the stage with their profile pictures and qualifications revolving on the cinematic projection screen behind them.

The reason the NANS is a good conference is because the American market for SCS makes up 50 percent of the world's total spinal cord stimulator revenue, with the remainder of the first world making up the rest; therefore anything that is new or interesting is likely to be presented at the US meeting. It was at this conference that our pain clinic started its journey with paresthesia-independent SCS because it was where the author of the study presented the preliminary results of the randomized controlled trial on high-frequency SCS versus paresthesia-based stimulation. After hearing the presentation I decided that I should give more consideration to his therapy. The scientific paper of that presentation was published later that year.

Interestingly, at that meeting were a group of people promoting a new spinal cord stimulator therapy, but the US Food and Drug Administration prohibited them from speaking to American doctors about a therapy it had not approved. These representatives were in full view of anyone passing by but could not engage with any American doctors, and there were even monitors there to check that there was no interaction. Weird.

Now that we have entered the age of paresthesia-free therapies there has been a mammoth gunfight to compete with the 10 kHz system. There have even been court battles over patenting a frequency. Papers have been written to undermine 10 kHz as an effective therapy. Trials have been hastily put together to show that 10 kHz is no better than 1 kHz and that the therapy consists largely of a placebo response. Mud has been slung and evidence interrogated. At the end of my talk at a meeting the other day, one of the attendees, a very experienced implanter, said to me, in the open forum, "I'm very happy for you, but remember, these guys"—the company whose therapy I was talking about—"are bastards. There's a reason why their share price is dropping, and you will find that their batteries don't last very long." My response to him was that based on our implementation of the therapy and unceasing collecting of data, as well as the published evidence, all I could do is hope that the company is acting in good faith. He came up to me privately after the meeting to apologize for his remarks and wished me well. I was interested to note that his main preoccupation appeared to be with the share price of the company, and I was later told that he had invested quite heavily in the company and had lost money accordingly thanks to the vagaries of the stock exchange.

There is a lot of money to be made from implanting devices—and from persistent pain and generally in medicine. Chronic pain is a global problem, and as I mentioned earlier in the chapter, neuropathic pain affects 6 to 8 percent of people within any given population. For a long time paresthesia-based therapies dominated the spinal cord stimulator market, and the competition between companies was all about how slick their programming was in the operating room and after the implant was inserted permanently, so that patients got the tingling in the right place. There was a huge competition in battery size, how quickly it would charge, and how long it would last; some manufacturers had batteries with definite failure dates, guaranteeing the need for a replacement. At the mercy of all these shenanigans is the patient. Companies fight among themselves to the detriment of therapies that have a good evidence base, and clinicians fight among themselves to the detriment of patients. Government organizations that rely on clinicians to provide them with objective opinions are left confounded.

The past ten years has been a constant struggle for me in the world of neuromodulation. I have had to struggle to reform the unit that I work in to implement a robust quality-assurance framework in infection control policies and operational policies while continuing to develop my own skills. I have had to get all our psychologists on board to situate the therapy within the context of a disability- and stress-reduction model, and I have had to train another colleague in inserting these devices. I have had very little help from the world of neuromodulation. I have found that those who are the most experienced in implanting are more concerned with jumping on the next research bandwagon, in order to improve their résumés and line their pockets or inflate their egos, than with ensuring that patients have greater access to the therapy and that the next generation of implanters is trained and appropriately supported. I have heard stories about clinicians in senior education positions within pain medicine who continue to use paresthesia-based therapies because they feel that the skill of an implanter is in their ability to place the lead, ensuring that paresthesia is produced that covers the patient's entire pain area—not in placing a lead over a defined vertebral level. This attitude is held and expounded despite the high-quality evidence demonstrating that paresthesia-free therapies are superior.

The evidence is the evidence—not at all perfect, but in the words of the philosopher Karl Popper (1902–1994), "Our aim as scientists is objective truth; more truth, more interesting truth, more intelligible truth. We cannot reasonably aim at certainty. . . . Since we can never know anything for sure, it is simply not worth searching for certainty; but it is well worth searching for truth; and we do this chiefly by searching for mistakes, so that we can correct them."

I think it is churlish and disingenuous to demand rigorous scientific studies, which a randomized controlled trial aims to be, in order to prove objectively that one therapy is superior to another or has an effect over and above a placebo response and then to turn around and say that actually the "in my hands" argument has greater weight. The expression "in my hands" is used by doctors who believe that, despite the results of scientific studies that have tried to reduce bias, their personal experience of implanting thirty to fifty patients and collecting (oftentimes retrospective) data supersedes the controlled trial, with its attempts to minimize bias.

We still use paresthesia-based stimulation in our unit, alongside the newer treatment, because we believe that stimulation therapies are iterative, and so we offer patients the option of whichever device we believe they should trial based on an assessment of their clinical problem. What is important to patients includes not only pain relief but also the impact on their daily life, such as whether they will be able to drive with a therapy or sleep with it on.

The specialty of pain medicine is in crisis. In a time when the provision of interventional pain therapies is being rationed because their evidence base is so poor, neuromodulation in the form of SCS is probably the best evidence-based interventional therapy, but it too is not always available for patients. Due to the rationing as well as the closing of units that provide only interventional services, we are losing people and not attracting a new generation to the specialty. Neurosurgeons are generally not interested in functional neurosurgery in the form of implanting spinal cord stimulators, so a therapy with a robust evidence base and life-changing outcome is not being supported.

A therapy that is not lifesaving but aims to improve quality of life is easily sidelined if it is not championed. This is less of an issue in a society where healthcare is provided privately and is driven by the incentive for doctors to provide therapies in order to earn a living. But in a publicly funded healthcare system where we depend on the government to decide which therapies are offered and where we are not taking responsibility for the consistent and national delivery of evidence-based therapies, we are very vulnerable. The challenge in private healthcare systems, on the other hand, is to curtail the enthusiasm of doctors when it comes to trialing patients. The trial–to–successful implantation ratio in some of these privately funded systems, for example, is around 40 percent, whereas in the UK's public sector it tends to run between 70 and 90 percent. It is not hard to see that in a private system, where you get paid for every patient you trial, there is a financial incentive to trial patients who might not be suitable.

Despite greater evidence than we have for any of the rehabilitation therapies we offer, we have colleagues who continue to think of SCS as voodoo. For various reasons, they refuse to accept the difference that this form of pain relief can make to a patient's life. The traditional view from behaviorism is that in chronic pain management the aim is not to reduce pain but to improve function. This mantra leads some clinicians to abandon the pursuit of pain relief. Ultimately, however, it is merely two sides of the same coin, because both SCS and rehabilitation therapies aim to modulate the nervous system. Whether neuromodulation is brought about by drugs, electricity, or talking, the outcome for the patient is the same: an improvement in quality of life and an improvement in the individual's ability to serve their community and their family.

Socrates is credited with having said "The only thing I know is that I know nothing." As I get older I find that this is the only certainty in the practice of medicine, and in life generally.

CHAPTER NINE

Do Not Go Gentle into That Good Night

Getting older is of course inevitable, and for some the aches and pains and changes in the body are simply perceived as a normal part of aging. Gradually people contract their world as their ability to function diminishes. My father and I were having a discussion along these lines this morning. I asked him how he was feeling, and he said that he was feeling "subpar." My dad is seventy-three, although he rounds his age up to seventy-five when he thinks we are not taking him seriously. He has acquired the diseases of a well-lived life and is on the medications that accompany these maladies. He is restricted in taking some medications because they can interact with his other prescribed drugs, and he must avoid certain foods that can elevate his blood sugar or precipitate an attack of gout. He does push the envelope when it comes to his gout and diabetes, and sometimes the temptation of meat or an ice-cream float is too much.

He said that in his office that morning he had seen a few seventy-three-year-olds who were slipping gradually into dementia. He was reflecting on why some people in their seventies appear to be able to soldier on, while others—and here he used an Afrikaans word—are *inmekaar*, which literally translates to "folded in on themselves." I mused that in my experience individuals who continue to engage in work, keeping their brain from atrophying, and who remain socially engaged are less likely to experience this gradual decline. I also told him that people who continue to exercise and use their muscles are less likely to experience physical atrophy. I have noticed that some people give up on living as they get older, and I think that this programs their cells to start committing ritual suicide. I told my father that the reason he has not declined into dementia and is not *inmekaar* is probably because he has continued to work as a GP. (I later heard from my mom that he was feeling subpar because his gout was acting up.)

The elderly patients I see in the UK tend to fall into one of two groups: those who are aggressively pursuing health by regularly going to the gym and spending a large part of the day doing mental puzzles, and a contrasting group who appear to be waiting before St. Peter's gate on a permanent basis. These two groups seek healthcare for very different purposes: the former group is looking for functional improvement, and the latter is looking for palliation and comfort. For those who choose to *not* go gentle into that good night, this gradual decline associated with abnormal feedback from their body is unacceptable, so they sally forth seeking a cure. Intolerance with inevitable aging has resulted in an explosion of the medicalization of aging and aggressive marketing of wellness. (In South Africa there is a chain of health shops called the Wellness Warehouse.) There is an adage: We ruin our health in search of wealth, to seek, to have, to save, and then we spend our wealth in search of health only to find the grave.

Older adults like my father live on the edge when it comes to health and well-being. Failure of organ systems, including heart failure and kidney disease, as well as diseases like chronic obstructive pulmonary disease, may result in the experience of chronic pain, and drugs used to alleviate this pain may not work as well or their use may result in unanticipated side effects that make the consequences of treating the condition harder to bear than the condition itself. The older person's nervous system changes, and this has implications for the experience of pain. The fatty layer of the nerves in the spinal cord is reduced and there are fewer connections within the brain, particularly in those areas involved in the processing of information about tissue damage. Age also brings with it a decrease in the body's naturally occurring opioids, which help regulate pain without any intervention. However, it is unclear whether these physiological alterations cause a difference in the pain response of the older person to harmful stimuli. They may offer some evidence of very minor increases in the ability of the older person to resist mild pain as they get older, or this evidence may merely reflect greater stoicism by the older adult (or fear of an "I told you so" from their children when gout visits) and their belief that pain and discomfort are an inevitable consequence of advancing years.

What we do know is that older patients report less pain, or pain in a different distribution, compared to young adults with the same clinical problem. Pain assessment may be difficult in the older person because of changes in cognition along with reduced memory and an inability or incapacity to report pain due to dementia; we have therefore developed specific pain scales that rely on factors other than self-report to assess the experience of pain. Restlessness, grimacing, or sounds such as grunting and groaning may point to an individual's being in pain. These nonspecific behaviors are difficult to make sense of, however, and managing the pain of an individual who is unable to report what they are feeling is challenging. The struggle is to avoid giving patients aggressive quantities of usually opiate-based analgesics, rendering them more vulnerable to the complications of these medications, rather than grappling with the fear that we are leaving somebody to suffer.

Chronic pain has a significant effect on patients who are older. It results in reduced mobility and avoidance of activities and may cause individuals to fall due to their inability to rapidly change position. The consequent impact on the individual's ability to function may lead to depression and anxiety. Family and social relationships are disrupted, and chronic pain in older adults places a significant burden on the communities within which they live. Increasing social isolation and a reduced capacity for travel impact on the ability of older adults to access services, and stoicism and not wanting to bother the doctor also play a role. One of the most pressing challenges in the management of older adults with chronic pain is access to rehabilitation therapies in the form of behavioral interventions and physical reactivation and conditioning.

Spouses have always played a significant role in the management of patients with persistent pain. Their role is fundamental in the delivery of emotional support because they are the patient's constant companion, as the following story of Alan and Louise illustrates.

> *Dear Abdul,*
>
> *Well I think after twelve years I can call you that—under the circumstances. I just want to thank you for all that you've done for Louise over the twelve years. As for no doubt she would have been in a wheelchair if it wasn't for you. I hope now that you are Professor Lalkhen—I hope that you won't stop seeing Louise as you know she will not let anyone else do her treatment. Well Professor I'm sorry I will not be seeing you again. I hope you will carry on doing the good work that you do and have a good life. Carry on taking care of your mum and dad like our Michael will do with Louise. I enclose a small gift to you which I hope you accept. With kind regards for all that you have done.*
>
> *Best wishes,*
> *Alan*

I first met Alan and Louise while I was still a trainee doctor and they were living in England. When I became a consultant, I continued to see

Louise and perform her four-monthly steroid lumbar and local anesthetic injections. After they had both retired and moved to North Wales, which is outside our catchment area, they would drive all that way together to see me for Louise's clinic review appointment. Louise is my patient, but over the years Alan became my friend.

I received this letter from Alan, delivered by Louise on the day of her next scheduled injection. Louise had phoned a few months earlier to let us know that Alan had been diagnosed with lung cancer and had been told that he had not long to live. She spoke to my secretary, who has been with me since I started as a consultant. (In fact I insisted—as much as one can in a publicly funded organization—that having Julie as my secretary was a condition of my employment.) Over the years Julie has grown close to Louise, even though they have never met in person, and both of us have always received Christmas gifts from Louise.

Twelve years is a long time to know anyone, and patients become part of the texture and color of life as a doctor and as a person without our even realizing it. Usually injections are performed every four to six months and review appointments are arranged perhaps once or twice a year. Each interaction represents a snapshot in the life of the person we are treating, and in the same way that comparing photographs taken months apart allows us to appreciate change more accurately, this pattern of interaction facilitates a clearer view of the inexorable progression of time and its impact on form and feeling than is often the case with people we see on a day-to-day basis. The snapshots tend to make more of an impression and the changes appear more dramatic. I see my parents who live in South Africa three or four times a year, and my anxiety about this infrequent interaction makes me more aware of the changes they are going through in the same way that I am mindful of how my patients are changing—not just in terms of their pain conditions but in the context of their overall aging process and the inevitable decline that is associated with this.

One of the reasons I decided to embrace pain medicine as a subspecialty of anesthesia was this yearning for a relationship longer than the five minutes it takes to ask about gastroesophageal reflux, the last meal eaten, and previous anesthetic history, which is usually the sum of an anesthesiologist's conversation with a patient before surgery. As doctors we sometimes feel that these interactions do not touch us and that patients pass through our hands without leaving an impression. But I remember unfolding Alan's handwritten note and reading the words and realizing how much he had come to mean to me. Even though it was Louise who was my patient, I always took the opportunity when seeing them to gently admonish Alan about his smoking. During the time I knew him he had treatment for coronary artery disease and suffered with numerous chest infections, but he never missed attending an appointment with his wife. Our common ground was cars; Alan found it amusing that I had moved from owning various cars to eventually giving up cars altogether and riding a bicycle. His final parting gift to me was a watch and a bicycle pump. I remember sitting down in my office and reading his note; I could almost hear his labored breathing and see his ready smile as he wrote it. I found myself unexpectedly crying as I sat at my desk.

I saw Louise some months later for her regular spine injection. Louise is extremely thin and has struggled with osteoporotic fractures of her back, which means that over time the bones of her spine have thinned out, and as a result they have collapsed under their own weight, changing the shape of her spine and resulting in an abnormal and painful curve in her upper back. These areas become painful because of inflammation between the joints of the spine. She had struggled with lower-back pain for several years, which I treated with steroid and local anesthetic injections. Over time the pain would flit from her lower back to her neck and then to the midsection of her back, where her fractures were. She would come and see me in between injections, and I would prescribe appropriate medication as well as general advice relating to movement, particularly activity pacing, which she always found challenging. Alan and I would talk about cars, his comments being either derisive or complimentary, depending on the type of car I was currently driving.

Louise would complain about Alan's failure to heed advice about his own health, and I would have to referee between the two of them, trying my best not to side with either.

I learned about love and tolerance and respect from Alan and Louise. They were like a well-rehearsed duo, finishing each other's sentences and seamlessly picking up each other's thoughts, producing a stream of loving consciousness. I would often say very little during the consultations, preferring to listen.

At the time of her injection following Alan's death Louise was wearing a formal black dress. She was still extremely upset, and I had asked the nurses to make sure she was placed in a side room while she waited. I remember sitting with her for a long time, reminiscing about Alan and his passing. She seemed even more gaunt than usual, sunken into herself, and appeared to be struggling to find the energy to carry on without him. It was almost as if by wearing the black dress she was clinging to the memory of her last contact with him, which would have been at his funeral. She had decided not to leave Wales and move in with her son, who lives closer to the hospital where I work, because at least in the village in Wales she has friends, whereas if she moved back to England, she would be alone all day. Her son now comes to appointments with her, taking the place of Alan. I last saw her just before Christmas, and she admitted that she was coping poorly without Alan and that this time of the year was pregnant with memories that threatened to overwhelm her. The pain of loss and loneliness may well be harder to bear than the pain from an aging body.

Bereavement and loss are ever-present companions in life, but more so with advancing age. I have now been a consultant for a decade, and many of my patients when I first started were already in their late sixties and seventies. They are now losing husbands and wives and sometimes children. Sometimes when Julie phones up to arrange their next injection, we are told that the patient has passed away.

The passing of the patient's loved ones impacts on their pain experience by changing their ability to suppress the abnormal sensations coming from their body and curbing their desire and motivation to carry on regardless. Their losses also impact on me. Sometimes it feels as if I am walking with a group of people whose numbers are ever diminishing. I tell Alan's story because I know he would have wanted me to communicate how challenging it is to support someone who suffers with persistent pain but also to leave a legacy of his great love for Louise. She wants me to tell his story because of her great love for him. It is these stories that make working in this field a privilege.

———————————

Grace is ninety-four and always wears a suit when she comes to clinic appointments. She is nearly doubled over but refuses to use a walking aid because she says that it makes her feel old. She is always apologetic, saying that my time could be better spent looking after older patients. Over time Grace's spine has remodeled, as spines generally tend to do. I explain to patients that when they are young (and I mean twenty years old or younger) their spines are like cars in a showroom. Over time and with use, the car's paintwork fades and the car bears the scars of its interaction with the environment. While the car continues to function, it no longer looks like that showroom version. Studies have shown that age-related changes occur in the spine beginning in our twenties, and these changes are described with words that inspire fear, such as "disc protrusion," "disc bulges," "degeneration," "nerve compression," and "facet joint degeneration." As I previously touched upon, however, these changes are also apparent in patients who have no pain from their spine area, leading doctors to conclude that none of us retains a showroom spine as we grow older and, frustratingly, that it is difficult to relate pain to changes found on an MRI scan.

Grace's spine is like a classic car that has not been lovingly cared for. She has age-related changes in the architecture of her neck, and the bones have remodeled and altered to the point where they are pressing on the collection of nerves called the cauda equina (or "horse's tail") that extends from the bottom part of the spinal cord. Surgery to correct the deformity in her lower back would be a massive undertaking, and because of all her complex medical problems she would probably not leave the hospital following such a procedure. Her main complaint is that her legs feel numb and she doesn't always know where they are in space and time. It is painful for her to walk, and this limits her ability to shop and do her chores. She cannot stand when she bakes, and she feels keenly this loss of an activity that made her the center of her social circle. She has tried many of the medications we normally use for nerve pain, but they have made her too sleepy or muddled or constipated.

Grace attends the clinic with her son and daughter, who are both in their late sixties. Her son has had knee replacements and walks with a cane. I find it poignant that her children are so old, but I suppose when you are ninety-four your children can easily be in their seventies. Her children are around the same age as my parents, and part of me hopes that I will be able to attend a clinic with my parents in the way that Grace's son and daughter do. She retains much of the fierceness that I suspect made her formidable in her youth. Looking back at the challenges she has faced in her life, it emerges that she struggled significantly with mental health problems in her thirties and forties. It is quite easy to forget that the same maladaptive patterns of behavior that patients experienced when they were younger follow them into old age. Anxiety and depression, which exacerbate the experience of pain, have the same effect when individuals are older as when they are younger. Sadly, what does change with age is the cognitive flexibility to solve problems, as well as the ability to access therapies that are more readily available to younger patients suffering with psychological disorders.

The consultation and conversation I have with Grace and her children mainly involve explaining why she has weakness in her legs as well as shooting pain. I explain that these pains are not dangerous and that the best thing to do is to try to maximize the level of activity she is able to do and to understand that the more frustrated and discouraged she becomes, the worse her pain will be. She finds some comfort in an explanation that allows her to manage her anxiety at what is becoming an ever-shrinking world. She tells me that she feels as if she is looking through a window and what she can see is becoming more and more faint. The explanation I give her will have to suffice in the absence of being able to offer her definitive treatment for her nerve pain and leg weakness. We—her daughter, her son, and I—talk to her about how helpful using a walker would be, but she simply tuts at us. Her son's eyes spark with frustration and resignation at a look that he has probably received from his mother all his life.

It is not always the case that we cannot offer older patients any form of interventional treatment. Grace's management contrasts with that of Maria, who is eighty-four and looks disconcertingly youthful, so that I sometimes have to remind myself of her age. She is a small woman, always smartly dressed, with a beautifully cut black bob hairstyle, and she always wears full makeup. Maria has nerve pain in her right leg because the outlet of one of the nerves in the bottom of her spine has become gradually narrowed. I treat her nerve compression with a steroid injection around this nerve every six months, which allows her to continue to live independently. She lives alone and does all her own housework, including arranging the ends of the Persian carpets on her floors so that they all line up. She travels across the country frequently and has an active social circle. The six-monthly injections to "discipline" her "badly behaved" nerve, as she calls it, are her lifeline, she says.

The point of these stories is that effective pain management must be tailored to the individual and be flexible around them. The stories also highlight that small interventions have the ability to translate into significant gains in independence.

═══════════

The average age of patients I see for repeat injection of local anesthetic and steroid injections for lower-back pain is seventy-seven. While there may be no relationship between the way a patient's spine looks on an MRI scan and the presence of back pain, nerve pain radiating down an arm or a leg is a different matter. There may be a correlation between what is seen on a scan and the sufferer's pain, which usually occurs within the normal anatomical distribution of that nerve. Spines remodel over the years, and new bone is laid down in places where bone should not be created; nerve canals that were once capacious become narrow and impinge on nerves. The problem, however, is not always mechanical, and surgery to relieve the pressure around the nerves does not always result in a pain-free patient. In these cases it may be that the nerve has already been damaged by the pressure exerted on it, like a peach that is compressed; while the skin of the peach may look unbroken, the flesh below has been bruised and damaged.

I often explain to patients that while nerves look like cables from the outside, internally they are made up of millions of fragile tubes that are easily disrupted. This narrowing, which is called spinal stenosis, occurs in the lumbar region; compression of the nerves within this canal acts like turning off the flow of water in a dam so that less water is reaching the periphery. The compression is dynamic, and patients describe being able to walk a certain distance before the narrowing becomes so severe that the flow of water, as it were, ceases altogether, causing them to feel weakness, but more often pain, in their legs. They then must rest a while to allow the swelling to reduce, the water to begin to flow again, and then they can carry on walking. This is their "claudication distance." Patients describe how they can no longer perform activities at their previous pace. The more active somebody was, the more intolerant they tend to be of this kind of limiting pain and consequent loss of function. Injection of a steroid into the nerve canal sometimes relieves the pain, but the evidence for this therapy is poor and it is no longer recommended by NICE in the UK. There are patients, however, who are intolerant of medication and are unsuitable for surgery for whom this injection is the only therapy that will help them to function.

Moira is suffering from this narrowing of the spinal canal and is ninety years old, only about four feet tall, and has come to the clinic today with her daughter, who is seventy, and her grandson, who is probably nearly forty. She has a reduction in the diameter of the spinal column containing the spinal cord in her neck and the cord is being strangled like a river choked by overgrown vegetation along its banks. She constantly feels like she is holding grit in her hands because the nerve signals are not properly getting through to where they should be. She says she sometimes talks to her hands and tells them "Don't be so soft," while she tries to wipe this nonexistent grit away. Unfortunately she would not be able to tolerate a major operation on her neck and so has to live with this feeling of grittiness in her hands and an inability to hold anything properly. Her walking is also affected, but she says that she is too young for a walker. She is frustrated because she still lives alone and wants to do her own gardening; her mind and her body have failed to keep pace with one another. The family is quite protective and worries about her; she has been with them for so long that they cannot imagine a time when she would not be part of their lives.

She has tried lots of medications to manage her symptoms, all of which have made her hallucinate, and today we agree that she should take a low dose of codeine combined with paracetamol. She is initially quite reluctant when I say she can take up to eight tablets a day. After I have warned her of the possibility of constipation with this medication, she tells me that eating chocolate now gives her diarrhea. I respond by saying that if I ever reach ninety and can no longer eat chocolate because it gives me diarrhea, they can take me away. She laughs at this. I tell her to take the codeine regularly with three bars of chocolate.

Despite her pain and discomfort, together with her growing inability to self-care and be independent, she laughs. It amazes me how indomitable the human spirit can be in the face of adversity.

―――――――――

We don't really understand why human beings suffer with pain in the joints as they get older. Nonspecific lower-back pain is sometimes attributed to the changes that occur in the small joints between individual vertebrae as a person ages, a condition referred to as facet joint pain or facet joint arthropathy. The cartilage within joints does not have a nerve supply and therefore feels no sensation, and the belief is that the abnormal deformation of the joints is what results in the pain because the brain perceives the joint to be in danger; this probably accounts for why weight loss helps with chronic knee-joint pain, and the same may well be true of the lower back. The changes seen on X-rays and MRI scans of the facet joints and the knees have led to therapies that focus on these areas. The knee is a relatively simple target, and replacement joints have been created, but this is less practical in the spine. The equivalent treatment was once considered to be fusion surgery of the lumbar spine or replacement of the intervertebral disc. Both these procedures, however, have been shown to not improve chronic lower-back pain and have since been abandoned. The facet joints are most active when you lean backward, which has been used as a sign that the lower-back pain experienced by the individual is due to a pathological issue at the facet joint. Bending, sitting, standing, and twisting all become associated with pain. The change is gradual over many years, and the impact is felt with activities that are innocuous to anyone who does not have pain—changing the duvet cover can become a major hurdle in life. All of my patients are fiercely independent; despite the passage of time and the loss of loved ones they prefer to continue living by themselves rather than moving in with a relative. Often they are accompanied by caring and desperate children who want their parents to come and live with them, only to be met with the refusal to be a burden or a desire to continue to have their own space. The injection of local anesthetics and steroids into the facet joints of elderly patients is thus an important therapy, but because of the poor evidence base associated with this therapy it is becoming increasingly difficult to fund it.

There is no explanation for why patients report benefits from an injection of local anesthetic and steroid into a joint, when we know the pain is in fact due to changes with the function of the nervous system. But for some patients it does make a difference, providing a means of coping with pain; others say that it does not reduce the pain completely and that over a period of six months the effect gradually wanes. There are individuals within the scientific community who attest to the effect being due to placebo and the activation of descending inhibitory pathways triggered by the patient's belief in the therapy. Some argue that the effect of the steroid has a knock-on impact on the desensitization of the nervous system. I sometimes wonder if all these patients really benefit from is the continuity of care, knowing that a doctor will see them on a regular basis and will provide a safety net for what they perceive to be their inevitable and inexorable decline into needing to be looked after.

We are now restricted in terms of whom we can offer facet joint injections to. Failure of the medical community to establish exactly which patients benefit from this therapy has potentially resulted in the loss of a useful tool. This procedure has been used inappropriately in young patients, where the aim of pain management should instead be to rehabilitate the individual utilizing an approach that recognizes the distress and disability caused by chronic pain and attempts to reverse this situation through a process of cognitive behavior therapy combined with exercise. We have also failed older adults in combining these therapies with reactivation of muscles through the provision of activities designed to maintain and improve muscle function and joint flexibility, as this may not be suitable for those in their later years. For some patients, as they get older these injections literally save their lives.

Patients tend to come for their injections at the same time, and I often find them in the waiting room chatting in groups of three or four. Social circles are formed and friendships are made around their shared treatment and suffering. Someone will ask where So-and-so is if they are not at their usual injection slot, and sometimes, sadly, I must inform the friend that their fellow injection patient is no longer with us. The news is greeted with a resigned and sad inevitability and the unspoken knowledge that one day they themselves may well be spoken of in the same way.

In older people we have a repository of knowledge we often fail to access. I also find older patients more reflective about their pain and more understanding of the limitations of the science of suffering.

Most of my patients tend to be women simply because they live longer than men. I spend a lot of time comforting patients whose husband has died. There is no way to capture the depth of loss associated with a friendship that has lasted for more than five decades. When I speak to somebody who has lost a life partner of this duration it is as if too much water has been added to a watercolor; the person becomes indistinct and blurred, with none of the vibrancy they once had. They are still there in front of me, but the difference is palpable. The adjustment to no longer being in a partnership is difficult, to say the least, and is often associated with a significant exacerbation of their usual pain symptoms. After her husband died one of my patients got herself a dog and became convinced that the dog was the living reincarnation of her husband because many of the dog's mannerisms mimicked those of her husband.

I make no secret of the fact that the assessment and management of older patients is the part of my job that I enjoy the most. Providing pain management interventions in the form of injections, and sometimes more advanced therapies such as SCS, to the older adult is rewarding and met with a simple gratitude I find unceasingly humbling. Older patients are like ancient trees that have witnessed the passage of time, and like trees that have borne fruit, the branches are often lower. The lowering of the branches is due to an understanding of the fact that possessions and achievements pass and that what you are left with is the character you have built over time. This is usually not the case in younger patients who may be occupied with the challenges of child rearing and engaging in productive economic activity and for whom chronic pain becomes either a route to withdrawing from life and living on incapacity benefits or a life-long impediment to successful engagement with the world.

A day of seeing patients in their seventies and eighties for injections is a pleasure; I am always met with a courtesy and respect that I do not often enjoy in other aspects of my professional life. These patients are grateful for any small measure of improvement they receive from therapy. The older patient tends to dress up for their clinic appointment, and I often remark upon what they wear. I suppose they have wardrobes full of clothes that were once used for a work life. Older patients also tend not to answer their mobile phones during clinic appointments and often have stories to tell and offer a perspective on life that facilitates my own learning. There is less of a sense of entitlement and more of a sense of gratitude. During the appointment I see flashes of the personality that once was. Working with older adults in the context of a pain clinic has allowed me to realize that true value comes from how you treat other people and what you are able to give to others. My older patients take an active interest in the changes in my own life situation and always ask after my parents—probably because I frequently talk about them. My parents are going through some of the same changes that my patients are experiencing, and a bridge is formed between caring for the person in my clinic and looking back home.

Pain in older adults is a difficult and challenging societal problem. Chronic pain in the elderly results in an inability to function on a day-to-day basis. The signs of neglect are often there when one examines patients who are living by themselves. Meals are smaller and less complex and weight loss can be a significant issue. Loss of muscle results in an even greater reduction in the capacity to perform activities that require power. The world of the older person becomes smaller as their ability to walk is limited by pain. Vacations they once looked forward to are no longer possible because health conditions make obtaining travel insurance difficult. It is sad that persistent pain in the older adult is viewed by both providers and the patients themselves as an untreatable inevitability. Patients must cope with pain and the unavoidable loss of a spouse and function, as well as a society that ignores and marginalizes them while it values the shiny and new.

There is also the problem that, as previously discussed, medication for the management of pain is poorly tolerated with advancing age; there is a change in the sensitivity of receptors, and interactions with drugs used for other diseases of the older adult are common and sometimes catastrophic. The reality is that many of our medications do not offer much benefit but do impact significantly on the ability of an individual to think and function. In desperation to maintain their independence, patients are often put on opiate therapy, which causes constipation, addiction, dependence, and withdrawal, together with the suppression of the immune system and cognitive decline.

We walk a difficult line in anesthetics and in the practice of critical care medicine. Ours is not to judge whether prolonging life versus a natural death is the right choice. I once asked a surgeon who operates on people who have cancer of esophagus (the food pipe) what he would do if he was diagnosed with the condition. His response was "I would have a stent [a metal tube] placed into my food pipe to keep it open and I would go back home and drink myself to death." His response is an acknowledgment of the suffering some patients undergo.

In the Western world we work at the very edges of medicine, where the margins between quality of life and incurring suffering are blurred. I have found that in the United Kingdom the attitude toward medical intervention is very different than it is in South Africa. In South Africa if somebody has an illness that we don't have the ability to manage without heroic efforts, we tell the family that they should take their relative home and care for them while they are dying. This is met with acceptance. In the UK there is often the expectation that people will have every available therapy before shuffling off their mortal coil. It is difficult to know why levels of acceptance surrounding healthcare differ so much between places and cultures. It may reflect the differing availability of resources. It may be cultural, that a belief in an afterlife or ancestors facilitates greater acceptance of death. It may even be due to differing approaches to family structures; I have sometimes found that the relatives who push hardest for continued treatment appear to not have the best relationship with the patient. Perhaps they yearn for more time to make amends.

Sometimes I wonder how my own last days will be. As I watch some of my patients gradually fade away I wonder about the ethics of prolonging life at the expense of suffering. Pain management in older adults is complex, and these ethical conundrums cannot be avoided—although in the West we have a tendency to do just that. As our medical abilities and technologies continue to develop we are increasingly going to be faced with this dilemma, however, and it is crucial that we approach it from a well-considered and moral standpoint. I anesthetized a patient just yesterday with inoperable large bowel cancer for whom the surgeon was refashioning his stoma (the opening in his abdomen through which he now defecated). When he woke up from the anesthetic and I asked him if he was okay, his response was "No—why did you not just let me go?"

CHAPTER TEN

Show Me the Money: Private Pain Practice

We are such stuff as dreams are made on, and our little life is rounded with a sleep.
—William Shakespeare, *The Tempest*, Act 4, Scene 1

I visit Cape Town on a regular basis to see my parents and siblings; my brother is a GP and has a little boy who associates my visits with Lego, and my sister is an academic clinical psychologist. Over the past twenty years I have watched Cape Town change, becoming an international city where you can have anything your heart desires. It has become a playground for the rich and famous, with the advantages and disadvantages that accompany this. The city center and surrounding area from the slopes of Table Mountain to the Atlantic Boardwalk have been redeveloped and transformed into a bohemian, hipster, and coffee-lover's paradise. Areas that have been settled for centuries have now become gentrified, and those without money have slowly been squeezed out of neighborhoods like the Cape Malay area on the slopes of Signal Hill, moving farther and farther away from the playgrounds in the city.

The bakery where I like to have breakfast has a seating area where I can simultaneously have a view of Lion's Head and the city framed by the edge of Table Mountain. A short distance from this coffee shop is a very subdued and elegant building. A couple of Porsches and a Maserati are usually parked outside, and a sign lists as some of its occupants a cosmetic dentist, several plastic surgeons, skin-care rejuvenation specialists, and dietitians. The building also contains a specialized gym. The men and women who wander into these buildings are already quite beautiful. What I find very interesting about buildings like this (and I have seen them in many parts of the world) is that they very seldom contain the offices of clinical psychologists.

We live in a world of hashtags and Instagram streaks (the length of successive posts you are awarded "likes" for, as I was informed by one of my colleagues who has teenagers), and we have developed technologies to lift, puff, tuck, and freeze various part of our anatomy. On any given news feed the lives of people whose sole occupation seems to be their appearance take up the same amount of space as stories about corruption, food shortages, and war.

Even though we have centuries of spiritual and philosophical guidance that indicate that embracing the appropriate amount of sleep, correct diet, mindfulness as a virtue, and consistency of behavior is the most effective way to lead a long and contented life and ensure well-being, we aggressively and impatiently pursue technological solutions to our inevitable decline and place a premium on external experience.

Documentaries about centenarians from the island of Okinawa and in parts of Europe as well as in the US (often Mormons) all indicate that moderation, social cohesion, physical activity, and a sense of community are what facilitate longevity. Perhaps, though, we, much like James Dean, would rather crash and burn than fade away—we, however, tend to not have good-looking corpses. The pursuit of perfection and well-being makes us very vulnerable to promises of instantaneous fixes but also puts us at risk of a new type of pain.

The experience of pain is an ancient biological alarm system that warns us of real or perceived danger, telling us that we are about to be injured or that we have been injured, and is hardwired into the most primitive parts of our brain. But ultimately the perception of what is harmful is a matter of our own psychological disposition toward the threat or stimulus (having your bottom pinched by your husband or wife is different from a pinch by a random stranger—same stimulus, different threat interpretation) and therefore exists on a continuum. I imagine that the everyday feedback from the body of a professional rugby player, battered by their daily exercise regimen and high-intensity games, is not something that the rest of us would put up with because there would be no context for it. Not having a six-pack when we look down at our stomach does not leave most of us distressed, but for a magazine-cover model the experience would be truly frightening. This discomfort from our perception of what is and is not appropriate in terms of how our body feels or looks leads us to seek help. The manner and method by which we seek help are dictated and modified by the culture we come from and by the explanations and expectations that we have of illness.

In the twenty-first century, we have beautiful hospitals resplendent with technological marvels, including imaging devices that can delineate with stunning clarity the inner architecture of our body. We have sharp needles and processes to obtain blood from patients which is tested in laboratories using spectrophotometers. We have doctors and nurses and occupational therapists and physiotherapists—a veritable armada of people to manage healthcare. Hospitals have managers and accountants and marketing departments; they are like multifaceted temples or shopping malls for various conditions. In these vast complexes lies the vested interest of a host of people to maintain the idea that we, the healthcare professionals, are omniscient and omnipotent in helping you live well and live longer.

The often undisclosed reality, however, is that if you live in a society in which people have ceased clubbing each other over the head, if you live in a home that has a flushing toilet and you have access to soap and water on a regular basis as well as appropriate nutrition, you have probably already saved yourself from around 90 percent of the maladies that could affect you and for which you would need a hospital. If your car has airbags, you drive slowly, and you maintain the car, if everybody follows the rules of the road and observes safe stopping distances, then you have probably avoided most of the trauma that is not related to interpersonal violence. If you do not drink alcohol (the *Lancet* has now published evidence that there is no safe limit on alcohol) and therefore avoid subsequently falling down and fracturing your ribs, which accounts for a significant number of hospital admissions in our clinic, and you avoid sugar, which is the new fat, and exercise on a regular basis, then it is likely you need never darken the door of a hospital or your GP. If you love and live mindfully in the moment rather than living in the future, where you will become anxious, or in the past, which will make you depressed, and you maintain your interpersonal relationships and have a clear idea of how you want to live your life, whether that be in service to others or a spiritual belief that you are here to earn an afterlife, then it is likely that you will not trouble our psychologists. The only role left for healthcare then is genetic conditions over which you have no control and cancers that manifest as a result of your abnormal genes.

The best thing you can do to not end up in a medical institution is to preserve your psychological and musculoskeletal health. This is more easily said than done because we live in a hypocritical world where we tell people to eat less fat and sugar but shop in supermarkets that have aisles and aisles dedicated to chips and chocolate. We make fruit and vegetables horrendously expensive and we sell cheap, energy-rich, but nutritionally poor products such as refined bread and pasta, which those on low income have to rely on to bulk out their diets. We promote and denigrate in equal measure the meat industry and the dairy industry. We confuse everyone about whether we should or should not drink milk. Is dairy good for you or is dairy bad for you? Should you embrace veganism, or will this diet cause a vitamin B12 deficiency? Should you eat more plants? How much protein do you need? We do not make it easy for people to answer these questions. And of course, ultimately there is no diet that guarantees immortality. Even as we become increasingly obsessed with well-being there is a rise in people suffering from disabling chronic pain.

This growth in chronic pain is aided and abetted by sugary inactivity and nihilism. There is seldom talk of chronic pain as an epidemic, but that is what we are facing. There is little made of the role that a biopsychosocial approach to the management of chronic pain could play in our society, and no recognition of the work that should be done by healthcare providers in improving the overall health of the population, for example by reducing the consumption of sugar and encouraging a greater focus on exercise and mental well-being. On walking into a supermarket I often reflect on the paradox of our existence: as a society we have the science to support the huge benefits of not eating refined sugar, but we continue to sell it. We know we should exercise more and be active and yet have invented binge-watching box sets as a sport. What underlies these contradictions? Perhaps we don't consider ourselves worthy of saving, leading to apathy toward our own decline? Or perhaps it is more existential: because we have the ability to contemplate our mortality we adopt the YOLO approach (you only live once), wanting the easy things in life and disregarding anything that requires more effort or a longer-term investment. When we then inevitably break, we are not willing to take responsibility and instead adopt the "Fix me, Doctor" approach. In turn, doctors who earn their living by fixing you are not incentivized to help you change your lifestyle, which would cut off a source of their income.

The pursuit of well-being looks to medical intervention rather than a shift, both societal and individual, in behavior and lifestyle. This has inevitably led to an increase in private medical practice. I work in a public sector organization that is funded by taxpayers. I have therefore

been trained at the expense of taxpayers to offer to those who need them
therapies that are grounded in safety, appropriateness, fiscal neutrality,
and efficacy, and these therapies are free at the point of contact. I am,
however, restricted in what I can offer taxpayers by organizations such as
NICE and clinical commissioning groups made up of local GPs, which
assess which therapies are useful and demonstrate benefit based on the
available scientific evidence. This was not always the situation; just forty
years ago doctors were able to offer therapies that "in their hands" were
considered successful.

Archie Cochrane, a Scottish doctor, is considered the father of clinical epidemiology and evidence-based medicine; in 1979 he wrote, "It is
surely a great criticism of our profession that we have not organized a
critical summary, by speciality and subspecialty, adapted periodically, of
all randomized controlled trials." Since then, the autonomy of doctors to
make it up as they go along has been gradually curtailed through the
implementation of evidence-based practice, which is also sometimes
used to restrict therapies as populations explode and governments
become poorer. I have just received an email announcing that the local
commissioning group is stopping the provision of facet joint injections
for patients; overnight it has been decreed that we can no longer provide
a therapy that we have been offering for thirty years. There are cohorts of
patients who now will have to learn to adjust to a new reality with no
access to this therapy.

Not so if you have private healthcare, however.

If you are willing to and can pay, you will find a doctor who will be
able to offer you the procedure you are seeking. But this system too is
not without its flaws. I heard an orthopedic surgeon state during a
lecture he was delivering to trainee orthopedic surgeons that the threshold for offering a patient a view of the inside of their knee (an arthros-
copy) is based on the same symptoms but varies to a huge degree
depending on whether they are an NHS or private patient. The impa-
tience of the privately insured to get better rapidly, combined with the
fee earned by the surgeon, reduces the threshold at which intervention

is offered. Insurers who offer private healthcare are becoming more savvy to the discrepancies between public and private healthcare, ensuring that they are aware of what is available through the NHS and consequently restricting the services that they are willing to pay for, limiting these to something closer to what is provided to the general public. This is ostensibly in line with the evidence that dictates which services are available on the NHS, but it is also an excellent way of reducing what the patient is offered. I know of a pain medicine consultant who used to inject the patient's lumbar facet joints on their private insurance three times (which was the maximum allowed by the insurance company) before referring the patient to himself to continue the therapy within the NHS.

In the UK, when the NHS stops funding a procedure, some pain physicians simply offer that procedure in a different country, one with less regulation. Social media sites advertise doctors from the UK going to Dubai and to India, offering courses on X-ray- and ultrasound-guided injections, which are not funded by the NHS for chronic pain. These social media posts are usually accompanied by a photo of smiling individuals gathered in a lab, having learned a cadaveric approach to an interventional pain procedure. Included is talk about international collaboration to promote injection therapies for chronic pain. The fact that these interventions are no longer performed in the UK because they do not serve any purpose in improving function or reducing pain or disability is not mentioned. We are exporting our technical expertise but not our clinical experience. These unproven, even disproven, therapies are simply—for a price—shifted to countries that have started to experience their own epidemics of musculoskeletal problems related to living a first-world lifestyle.

—————

Private practice requires individuals to buy private health insurance, which is expensive. This has led to a divide between London and the rest of the UK, following the economic divide, with private practice far less prevalent outside of the capital. Consequently, private hospitals have not developed to the same degree, particularly outside London, and broadly speaking private medicine is not usually as well set up as NHS hospitals for major surgery. NHS patients are sent to a hospital specializing in the surgery they require, where they will have a tier of on-call health professionals consisting of trainees and, these days, a consultant. Our hospital has a consultant anesthesiologist available 24/7 to provide care to the sickest patients. We have fully staffed critical-care units and the laboratory facilities to manage these challenging cases. This is not usually the case in private hospitals in the UK. In South Africa the situation is very different; there the private sector is probably better developed than its public sector counterparts. There money talks, and it speaks very loudly.

A friend of mine told me that their mother in South Africa had been referred to a rheumatologist in the private sector; the next day she received lumbar facet joint injections for her back pain. Not only was the injection not supported by the scientific evidence, and not only was a thorough assessment of the patient's problem not made, but the intervention was performed using a CT scanner. This is a completely unnecessary, over-the-top intervention using an imaging technique that we would not use on a routine basis in the UK. But by performing the procedure in this way, the hospital can charge for the radiologist's time and thus help fund the purchase of that CT scanner. The aim of private hospitals and private medicine is to generate an income for the doctor and for the hospital providing the facility. Unfortunately, patients often approach private hospitals as if they were churches where they will be assured of benefit and where everyone is motivated purely by a desire to be of service. I am sure these hospitals and doctors aim to provide excellent healthcare, but ultimately there is a conflict of interest when an institution is being paid for every procedure, bandage, and needle used.

The patients I interview in the context of a medicolegal claim often reflect on their interaction with pain medicine consultants they have seen on a private basis. It becomes patently clear that the only way to make any money in private medicine is to offer patients a procedure. A conversation about chronic pain, an explanation that the pain is due to altered nerve function and that the aim of pain management is to help the patient start moving more despite continued pain, is not going to make any money. What makes money is offering the patient an injection of local anesthetic and steroid (every three months) or offering to burn their nerves for a small fee.

I have on many occasions considered taking on a private practice in pain medicine. We have surgeons at our hospital who offer private operations and who sometimes require a pain physician to manage those patients who go on to develop persistent pain. I have visited several of these hospitals, but I have never been able to see a way of replicating what I can offer through the NHS. From a financial point of view, it makes little sense for a private hospital to run a clinic that only gives advice, since the income generator is the delivery of a procedure. The remuneration from a clinic appointment does not cover the extra indemnity doctors have to pay. There is always the fear that in order to cover costs a clinical practice in the private sector will change ever so slightly compared to an NHS practice, becoming more invasive than would normally be the case in the NHS. I don't believe that I am an angel, immune to the lure of private practice, with its social media marketing and bevy of adoring, grateful, and passive middle-class patients, and I might therefore be tempted, albeit subconsciously, to be swayed in this way. So I have always vacillated when it comes to private practice.

I have tried to start a private practice within the setting of my hospital, but this has not been supported by the hospital management. This reflects one of the great difficulties of working in the NHS. Public sector organizations are not designed to aggressively increase revenue, unlike private hospitals. Hospital management does not therefore have the same approach to increasing and expanding services that a private sector establishment is geared toward. I am often frustrated when trying to implement anything new within the NHS. The first response is always "We have never done that. This is not how we do things." A private practice delivered alongside my NHS work would represent the best of both worlds. I would be able to give private patients more rapid access to NHS therapies that are not currently offered in private practice, such as neuromodulation and pain management programs, within the context of the usual safe and clean environment where I work. This would not disadvantage NHS patients, who would still receive these therapies after the usual wait, but it would increase the utilization of the operating room and clinic space out of hours, thereby increasing government revenue.

It could be argued that in the provision of healthcare everybody should have access to the same kinds of help based on need and how adversely they are being affected by a condition rather than because they can pay. One of the aspects of private pain medicine that has always distressed me is that those patients who are willing to pay for a consultation with a pain physician are often the most desperate. They are usually the most psychologically distressed, anxious, and depressed patients who catastrophize to a high degree about their pain. They are willing to pay to jump the queue in order to receive a magical cure for their persistent pain and will continue to throw money at the problem in order to resolve it. These patients usually benefit the least from interventional techniques but are the most likely to agitate for them. They are fish in a barrel when it comes to private practice. When I explain to these patients that there is no cure for their symptoms, they are the most likely to become angry, not just because of their condition but because of the money they have already spent. The temptation in that situation is to offer them an injection therapy in order to quell their anger and my resultant distress.

Private pain practice represents yet another one of these industries where people with too much money or sometimes too much desperation are willing to go. The doctor has a societal position that enables them to offer a therapy, but they can walk away from it when the treatment doesn't work. I know somebody who has had bilateral knee replacements but continues to be disabled because they have not lost any weight; they have no improvement in function because their underlying behavior has not changed. The person who performed the surgery and the person who anesthetized the patient, however, have benefited from this procedure, as the patient was in a position to pay for it to be done privately. I wonder about the ethics of saying to somebody, "I can offer you this procedure, but there is no guarantee that it will help you. If you still want it, then let's go ahead." At what point does our responsibility begin and end with regard to performing unnecessary surgery? Or is it the case that in a liberal society people can have whatever they ask for if they are willing to accept the risks? I suppose it speaks to a deeper question of whether we operate as individuals or as a global society. It strikes me that we are racing into oblivion as a species by not taking individual responsibility for our acts.

I have already alluded to my belief that pain medicine is an endangered specialty in the UK and that its practitioners are increasingly finding that the medical services they provide are being stripped away. Eventually, if we no longer have anything to offer a hospital that will generate revenue, the hospital will tell us to either go away or to do something that it does need. This is the reality of medical practice— perhaps more obvious in private practice but also evident in publicly provided services: that it is a business, with financial interests, and the patient is the product.

CONCLUSION

The Beginning of the End

Now this is not the end. It is not even the beginning of the end. But it is, perhaps, the end of the beginning.
—Winston Churchill, November 10, 1942

T his final chapter finds me on the Atlantic seaboard in Cape Town on a Sunday morning sitting directly across from Robben Island, once a prison, now a tourist attraction and a stark reminder of the inhumanity we can display toward one another, entrenched in law and justified by the selective interpretation of a religious text. It is a gray morning in Cape Town—Signal Hill and Table Mountain are still covered with stubborn clouds and are not yet visible, and the sea is more suited to being photographed in black and white than in technicolor.

The boardwalk that snakes from the Waterfront is a hive of physical and meditative activity. Some early risers stare out to sea, lost in thought, while others pound the pavement in colorful Lycra sometimes stretched beyond breaking point and sometimes embracing the athletic forms they were designed to serve. Some runners move lithely, in perfect rhythm, with bodies that display efficient power-to-weight ratios and the flexibility needed to chase their food, as if they were designed to run; others trundle along with minimal hip movement, knees groaning. All I can think about are the blank checks to hospitals and the future expansion of the bank balances of orthopedic surgeons, as this second set of runners slowly and, it appears, painfully grind away at the front part of their knees. Their dogs run around them, occasionally interested in their owner's progress and now and then staring up at them, wondering what the purpose of the run is because it clearly is not in pursuit of breakfast. Other Capetonians have decided that two wheels represent far better exercise and speed by in a blur of carbon fiber. The common thread is the pursuit of health and well-being; the form this takes is based on each individual's selective exposure to and interpretation of the information bombarding them.

If you have reached this point in the book, then you have seen through my eyes—sometimes clear and sometimes jaundiced but always, I hope, open and curious. This book about pain reflects the past twenty-two years of my journey and career, which started on the slopes of Devil's Peak at the University of Cape Town and moved to the far north of South Africa before a series of placements in the UK, in Plymouth, Warwick, Swindon, and Bath, finally ending up in Manchester and Salford. It is my journey of working in places where human life means little and interpersonal violence is rife, and places where extending life is of supreme importance—sometimes beyond what is reasonable, achievable, or kind—in hospitals where we barely had enough equipment and hospitals where scheduling lots of tests is the order of the day. This book reflects my personal journey, from not wanting to engage with people and preferring them to be anesthetized when I work with them to finding the greatest joy and satisfaction from long-term relationships with patients and their families.

Alongside my own story I have tried to provide an overview of the history of pain and our understanding of this phenomenon. What has become clear to me as I have written this book is that human beings have always been afflicted with pain—that we are creatures who are intimate with suffering and mindful of our decline and eventual end. We live such short lives and cower in the face of eternity, desperately engaged in immortality projects: children, money, fame, and the accumulation of stuff. This short life span gives us a unique but quite distorted perspective of ourselves and where we fit in to the span of human history; we rarely have the opportunity to witness change over even a hundred years. Perhaps that's why we repeat the same mistakes again and again—or perhaps it is because we cling to hope, left in Pandora's infamous box not as a boon but as a curse for humanity.

In a week's time it will have been twenty years since I graduated from medical school, and while I can remember elements of the past twenty years, and if you asked me to recall specific events I can, much of the detail is lost to me. The moments that are most memorable are those associated with the greatest pain, and not in a purely negative way—the greatest discomfort and times of discord have also been the greatest moments of learning and progress. We like to think that our choices are made in the fullness of joy and in moments of quiet reflection brought on by contentment, but the truth is that as a species we often make dramatic steps forward when we are most uncomfortable. We have a relationship with pain and suffering characterized not just by negativity but also by complexity and growth.

Pain has always been in the background, and sometimes in the foreground, of our evolution as a species. When we were single-cell organisms our only responses were those toward growth or away from pain. Even in our highly evolved state today, pain is still a teacher—the discomfort of emotional and physical adversity and injury prompts learning: "Don't touch that," "Don't go there," "Wear sunblock," "Don't date narcissists." We have fast pain fibers that get us out of trouble and slow pain fibers that result in any persistent unpleasantness laying down a memory of the event and teaching us to avoid that which caused us harm. Sometimes we accept discomfort and pain, viewing it as an experience that we need to endure in order to grow; we accept pain following a gym session based on the belief that our muscles will improve as a result. The report of pain always has context; in some situations inflicting harm is experienced as fifty shades of pleasure.

Perhaps our narrative of the experience of pain is also shaped by our technological ability to influence the experience. When there was nothing we could do to alleviate pain, we handled it by believing that pain was necessary. When treating pain became something that we actively and passionately sought to do because we understood the disease and knew of a treatment, then our enthusiasm knew no bounds and our philosophy became that it is inhumane not to manage pain. I wonder if this has less to do with pain and more to do with the structure and nature of our modern society, which values the individual, so that treating the pain of the individual has become an essential part of the human narrative.

Our technology continues to evolve, and our understanding of the neurobiology of pain evolves alongside the scientific tools increasingly available to study it. But we are still very much looking from the outside in, desperately feeling around to try to understand what this phenomenon is. What separates pain from many other symptoms is that it is so intimately tied in to how we interpret the world. Diarrhea is pretty much diarrhea; it has a beginning and an end, and if we can find a cause, then hopefully we can return our bowel function to normal. The nature of pain, however, is that it completely commands our attention—it is "aversive at threshold," to return to our opening definition. Pain activates brain areas that are primitive and archaic, present from when we were unicellular organisms; it sits in this ancient brain alongside the need for food, the compulsion to procreate, and the irresistible pull of sleep. Pain is a constant life companion, present not just physically but also psychologically and emotionally. Trying to understand and treat an experience like pain cannot be separated from trying to understand and manage people.

Having said this, ultimately understanding pain as a disease and then being able to target the molecular basis of pain is probably a better way forward than trying to ignore pain, distract ourselves from it, and therefore live well despite persistent pain. Pain is, after all, a primitive and hardwired alarm whose purpose is to warn that all is not well. Eventually we hope to be able to switch off the alarm system in the same way we hope to *cure* diabetes with stem cells rather than *managing* it by giving insulin. However, we do not currently have the understanding or capabilities to tackle pain in this way, and so we are left with trying to treat the experience of pain.

Our medical understanding is constantly progressing; today, for example, in many respects we understand cancers that once mystified us. Pain is sometimes (although not always) a manifestation of a cancerous condition, and in such cases we treat the underlying cancer in order to prolong life, arresting those cells that have gone rogue as best we can with instruments which in a hundred years will probably seem quite primitive. In doing so, we poison the abnormal cell, but we also taint the healthy cell alongside it. We irradiate both healthy and unhealthy cells and we cut out and brutally destroy the malfunctioning cancerous mass while trying to preserve the normal tissue. We understand the pain that accompanies cancer. We understand it because we believe it to be due to a disease that we comprehend. And yet patients with identical tumors can experience vastly different levels of pain, and we must therefore accept that even this apparently understood process is interpreted through an individual's unique cognitive, behavioral, and emotional framework.

We understand the immediate pain associated with damaged tissues. That is to say, we at least understand the source, but we do not always appreciate the variability. Understanding *why* somebody has pain is much easier than trying to understand how they feel and how they suffer. In seeking to provide medical treatment, as doctors we inflict pain daily—we damage tissues and then try to repair, reconstruct, and reformulate. We hope that in our destruction we achieve a measure of improvement—in function or in longevity—but pain is sometimes the inevitable consequence of that endeavor. We try to dampen it down for as long as is needed for tissues to heal and stop crying out against the insult they have endured. While doing this we try not to poison the patient with medications that may cause addiction and worsen their outcome or cause bleeding from their stomach or damage their nerves with needles that we place in order to deposit local anesthetic.

We struggle most when our current technology is unable to identify the source of pain. People with pain in a part of their body where modern technology is unable to identify anything that is broken or cancerous are viewed with suspicion. In the past we have consigned patients like this to asylums. About 60 percent of patients who were sent to asylums with seizure activity in the nineteenth century were found later to have brain tumors. More recently, we have tried to manage the patient's behavior by sedating them with opioids or damaging normal parts of their nervous system in order to quiet down the alarm that has been blaring. We have injected, electrocuted, tranquilized, and psychologized this group of patients because our technology is unable to diagnose the dysfunction that is causing them to cry out in pain. These patients are generally referred to by doctors as "heart-sink patients." They are caught in a vortex of misunderstanding and are not helped by a system that is designed to support the management of specific, identifiable conditions. We find it hard to accept that there are patients to whom we have to say, "I have done all the tests that we currently have at our disposal. These tests have not shown me anything in your organs that I can cure. We do not yet understand why some people develop conditions like yours. There are treatments, but they are designed to be empiric and there is no evidence that they will improve your function or reduce your pain, and there is a high probability that they will turn you into an addict or will damage parts of your body that are functioning normally. You are not mad, and you are not imagining the symptoms—it is just that we do not yet have the ability to diagnose your condition with a scan or a blood test."

One of the most difficult emotional situations that doctors find themselves in is when they have to communicate such explanations.

When we talk about a pain that science cannot measure, most patients end up believing that we are telling them the problem is all in their heads. I've been asked "Why is it that we can put a man on the moon but we cannot cure my pain?" more than once in my career. The fact is that we do not have the ability to analyze nervous system function at any more than an electrophysiological level, which measures gross conduction of electrical activity in nerves. Our ability to image the nervous system is confined to an MRI scan, and while we do have some tests that can measure blood flow in the brain in response to different stimuli, these tests are not yet specific enough to accurately diagnose a cause for many pain conditions.

———————————

I used to think very narrowly about pain, viewing it through the lens of a subspecialist, just managing the person's pain. I realize now that people exist in a context. Patients do not come to the pain clinic with just their pain. They may talk about only their pain, but they come with many psychological difficulties related to other life events and sometimes horrific social circumstances that I am completely unaware of until I read the letter from the psychologist. Patients are usually but not always overweight due to a poor diet; sometimes this is because of a lack of education, or they may eat to comfort themselves and eating becomes a way of coping. Patients may overuse alcohol as a means of coping, and may use prescription opiates in a similar manner; opiates become their new vodka. The aim with these patients is not so much to treat the pain as to treat the associated distress and despair, which in some way ameliorates the pain.

When I talk to patients in the clinic now I am much more aware of their hidden contexts and shadow selves. I explain pain to them as best I can and try to encourage an understanding of chronic pain. I don't want people to leave my clinic thinking that the pain is all in their head, but I do want them to understand that they have the ability to influence the abnormal sensations they are experiencing. I try to help them understand that every aspect of their lives influences their pain, and that is what we need to tackle. Their diet, for instance, is a manifestation of their distress but also contributes to perpetuating the condition. We often focus on the external manifestation of a behavioral problem or a maladaptive thought pattern more than we do on the actual thought or behavior, but both need addressing simultaneously. As I commented in the previous chapter, there were no psychologists in the beautiful gray building under the mountain. If there was a psychologist on the first floor, whom you had to see first, then it is likely you would progress to the floors above only if there was an identifiable physiological problem; you would see the dentist only if your teeth required treating, and you would see the skin-care specialist only if you actually had a dermatological condition. It is unlikely that you would ever make it to the floor occupied by the cosmetic surgeons.

As part of my clinic I therefore explain the current dietary recommendations that should be followed, as well as discussing how psychological wellness is predicated on a clear understanding of thoughts and feelings and that people need to find a reason for living and a meaning to their existence. It might be argued that I am exceeding my scope of practice as a pain clinician, but my view is that we shouldn't really be called a pain clinic; we should be called "a well-being clinic." It strikes me that all doctors should have this approach to a patient's wellness. Rather than focusing on inserting another stent into their coronary arteries we should reflect on why, in the three years since we put the last stent in, the patient has not made any changes to modify their risk factors. By not being this sort of doctor, are we not culpable of continuing our own middle-class pursuit of healthcare?

I have not even mentioned smoking or recreational drug taking because to me these pursuits seem so obviously harmful, but the very fact that these substances still exist and are readily available speaks volumes about our paradoxical attitudes to our own health. We continue to promote cigarettes even though we say right on the package that smoking kills. We bemoan the impact of soft drinks on the health of the population, particularly those on lower incomes, but we continue to manufacture these substances. We of course also eschew and decry wars all over the world but continue to produce armaments. I often reflect quite sadly on the fact that the many areas of the world where there is conflict do not themselves actually have factories producing the weapons; I then wonder whether, if the countries manufacturing them stopped providing guns, eventually people would get tired of fighting with sticks and stones. Perhaps because we are the only species that is able to understand that we will die, we sometimes in our despair rush toward self-destruction.

Pain medicine is not a pursuit that I would necessarily choose to return to in another life. Modern medicine is pretty good at treating injuries, and I often fantasize that if I were to come back as a healthcare professional in the next life I would be an orthopedic surgeon treating trauma victims. The bone is broken, and I can fix it—this is a simple architectural interaction with the patient. Disruption or discontinuity to the normal anatomical structures because they are physically traumatized is something we can easily repair, theoretically. Pain, on the other hand, is a complex experience and not one over which we have mastery, neither when it has been triggered due to injury nor when it has become chronic pain. In the acute setting we can manage the underlying or precipitating condition. We can remove the thumbtack from your foot and clean the wound and apply a bandage. We can put plates and screws on your broken bone and make sure the tissues do not become infected. We can treat your burns to a certain extent by putting dressings on them or taking away the dead tissue. We can treat your chemical injury appropriately. We can give you medications in the form of anti-inflammatories or local anesthetics or morphine until the nervous system's alarm settles down.

If you have persistent pain all over your body because of an accumulated lifetime of psychological and immunological injury caused by adopting a reckless lifestyle or live in a psychosocially poor environment, we do not have a medical cure. If you have developed persistent pain because you are too heavy for your knees—this is where we struggle. If your damaged nerve has healed but the pain has not resolved, we are less adept at managing your pain. Even therapies such as SCS, which today appear to be quite sophisticated, may be considered barbaric a hundred years from now.

I have become increasingly despondent over the past few years about my role as a healthcare professional. I sometimes feel that I am colluding with my colleagues, caught in a system that operates on the margins, with an understanding of pain that is no better than we had a hundred years ago. Pain medicine is one of the specialties that has created an entire industry based on the most inexcusably weak evidence. We are now being found out, and government bodies responsible for allocating healthcare resources are uncovering the waste of resources that have been plowed into injection therapies that have not been appropriately studied.

The most powerful and therapeutic part of my role as a healthcare professional is to explain pain. It is the only tool at my disposal that, in my experience, makes a difference to patients. Even when patients are angry and disappointed when informed that their condition is permanent and that the alarm system will never return to normal, the mere introduction to an approach acknowledging that there is no cure is in itself the start of a healing process. Most patients will continue on the roundabout of medical investigations and treatment, but at least at some point somebody has said to them "This is a permanent condition for which we do not yet have the solution." Unfortunately, as an industry healthcare can take a glib view of pain, and most people are never told by a physician about the complexity of pain. In most medical schools pain is given mere lip service as an experience. It is simply considered an introduction to the more interesting diagnosis using our latest technology and a very short period of foreplay before the main event of treatment.

Healthcare and medicine require a fundamental rethinking of where we are heading. Archie Cochrane asked us to collect and systematize the evidence for what we do; this has been done in many specialties, but there is still a deep need to implement, study, and improve on these therapies. Independent bodies are needed to look at where we currently are in terms of treatment and to design rational therapies without collusion with industry. (My dream is to trial a spinal cord stimulator system manufactured without a brand name, one that is able to offer all stimulation parameters within a single system.) We need to accept that many aspects of modern life are damaging to people, and that merely providing passive healthcare solutions is at best futile and at worst unethical. We understand how to prevent a huge number of diseases but invest very little in actually doing so because, ultimately, we profit from the managing of conditions that people inflict upon themselves, aided and abetted by society's marketing of the modern lifestyle.

Being a doctor today is complex, and I often feel inadequate in my role, with the pills and potions at my disposal often hailing from ancient times and simply repackaged for the twenty-first century. I feel that we have lost our way in medicine and are desperately searching for a way back in order to justify our worth. In the UK we have few infectious diseases, cars travel relatively slowly, and gun ownership is almost non-existent; I treat diseases of lifestyle and the ravages of time. Why are we not more involved in the primary prevention of disease?

Medicine continues to be a middle-class pursuit, and medical students do not reflect the wider society. There is also the concern that, over time, those who enter medicine become more interested in maintaining their role than in changing society. While it is not always the case, holding on to a perceived degree of wealth and position in society is what motivates some doctors. Combine this motivation with the rewards available in private practice, and in certain healthcare systems around the world people are exploited and subjected to unnecessary evaluations, tests, and procedures. Patients sometimes actively collude with doctors, accepting a significant psychological payoff for adopting the sick role, again facilitated by our modern society.

Who decides whether somebody should have a procedure depends on the country they are in. Ideally an independent institution, without affiliations to pharmaceutical and technology companies or anything that might lead to their own enrichment, should decide what is offered and, more important, what requires further study, based on the available evidence. Even in a publicly funded system such as the NHS, however, this does not always happen because of the inherent flaws in many scientific studies and because of lobbying by physicians who are subtly (and not so subtly) sponsored by drug and technology companies.

There also needs to be a shift away from doctors as the primary providers of healthcare and well-being and toward a more global appreciation of the factors that influence health, taking the definition of health put forward by the United Nations World Health Organization as a complete state of mental and physical well-being and not merely the absence of disease. The reality is that this requires an abandoning of self-interest on the part of doctors who should, first, do no harm; those elected to govern should define with evidence the policies that will facilitate well-being; and industries should be rewarded for providing therapies that are demonstrated to enhance quality of life. Given the current political climate and the general lack of global cooperation, I am pessimistic about this happening.

One of the great weaknesses in healthcare is that research is difficult to perform and is encumbered by so much red tape that most doctors in clinical medicine are put off from participating. It is a crime that we clinicians are not more actively involved in research, but we are so busy offering treatments (which do not always ensure a good outcome) that we are not given the time and space to ask the important questions that need to be asked about the therapies we offer. Governments are interested in telling patients that they have provided X number of procedures and X number of weeks reduction in waiting time. What is less sexy is being told that governments are investing more in research in order to answer the fundamental questions about what underpins good health. In the UK we can't even organize a meaningful sugar tax, and our attempts to reduce alcohol and tobacco consumption are pitiful, even though we know that these recreational activities end up costing society enormous sums—both physically and financially.

Continuing to work in a unit that offers pain management physiotherapy and psychology in order to reduce the distress and disability patients experience due to pain seems the best and least harmful option for me for now. Spinal cord stimulation for the management of peripheral neuropathic pain based on strong evidence is a worthy pursuit. Ongoing exploration of the dysfunction that is chronic pain and designing specific therapies to modify this system are global endeavors. Resisting the empiric use of cannabis oil or cannabis-derived products until clear evidence is established by randomized controlled trials utilizing these substances in specific pain conditions is a sure way to prevent this drug from becoming the newest method to medicate despair. Treating existential crises with alcohol, tobacco, cocaine, and methamphetamines will clearly not lead to an improved quality of life for the patient. Judicious, kind, and measured use of opioids, while communicating the risks and benefits of these drugs to the patient, is just. The most important work, however, is to explain pain kindly, gently, and plainly, based on our current understanding of what pain is and what the limits of treatment are. Highlighting the aspects of positive living, as found in societies where people live well and for a long time, is essential: whole foods, regular physical activity, living in community with one another, and having a shared ideology—finding your *ikigai*, your reason for being.

My hope is that after reading this book you will take responsibility for your health and ask questions of those from whom you seek advice when you are unwell. That when you are in pain you will be able to reflect on the screaming of your primitive brain and understand that the experience of pain is a biological alarm, but your response is a choice. Hopefully you will understand that pain can never be completely extinguished with medicines and can appreciate their potential for harm. Perhaps you will now be able to work with your doctor or nurse when you are in pain. If you have chronic pain, then I hope this book makes you feel acknowledged. If you know someone who suffers with chronic pain, then hopefully you are more informed, which is the soil wherein compassion grows.

In the words of the American physician Edward Trudeau, "Sometimes to cure, often to relieve, always to comfort."

ACKNOWLEDGMENTS

From my patients I have been taught the meaning of these words from Maya Angelou: "I've learned that people will forget what you said, people will forget what you did, but people will never forget how you made them feel." Thank you for the privilege of being able to serve you. This book reflects the skill and patience of the people I work with at the Manchester and Salford Pain Centre who have taught me and whom I continue to learn from. I would like to thank Dr. Tom Mount, who put me in touch with Ben Clark from the Soho Agency because he thought I might have something useful to say about pain. Thank you, Ben, for your guidance and encouragement. Thank you to Katherine Ailes for your thoughtful and kind editing of the manuscript, as well as James Nightingale from Atlantic Books for shaping the finished product.

INDEX

ABOUT THE AUTHOR

Dr. Abdul-Ghaaliq Lalkhen has been working in pain management for more than ten years. He is a member of the Faculty of Pain Medicine affiliated with the Royal College of Anaesthetists and is a visiting professor at Manchester Metropolitan University. He lives in Manchester, England.